D1200388

Riding the Waves

Riding the Waves
A Life in Sound, Science, and Industry

Leo Beranek

The MIT Press
Cambridge, Massachusetts
London, England

For information about special quantity discounts, please email special_sales@mitpress.mit.edu

This book was set in Stone and Stone Sans by Graphic Composition, Inc.

Printed and bound in the United States of America.

Library of Congress Cataloging-in-Publication Data

Beranek, Leo Leroy, 1914–
   Riding the waves: a life in sound, science, and industry / Leo Beranek.
     p.   cm.
   ISBN 978-0-262-02629-1 (hardcover: alk. paper)
   1. Beranek, Leo Leroy, 1914– 2. Acoustical engineering—United States—Biog-raphy. 3. Music-halls. I. Title.
TA140.B385A3   2008
620.2092—dc22
[B]
                                                                            2007013764

10 9 8 7 6 5 4 3 2 1

Contents

Prologue

September 23, 1962. Summer glides into autumn, topcoats appear, and New Yorkers brace for the opening of Philharmonic Hall, the first in the complex of buildings to rise at Lincoln Center for the Performing Arts on Manhattan's Upper West Side.

Scores of chauffeur-driven limos jam the intersection at Columbus and Amsterdam. "There's Jackie!" someone shouts. The First Lady, elegant as always, smiles and waves as she passes through the hall entrance. She has made a special effort to come. Back in Washington, JFK awaits the arrival of President Ayub Khan of Pakistan, a key U.S. partner in the ever-deepening Cold War. Jackie, the consummate hostess, will meet up with them later, but not before joining other luminaries here—John D. Rockefeller III, Adlai Stevenson, Dean Rusk, U Thant, and a host of others—to celebrate the start of a new epoch in the performing arts.

And here I am, too, a nerve-wracked soul among more carefree celebrants. I'm decked out like the rest in white tie and tails. I smile, but know full well that my role as acoustics consultant to architect Max Abramovitz will make or break my reputation.

The job has been a complex one, further complicated by politics, miscommunication, and poor choices. After months of travel to the world's concert halls and of interviews with renowned conductors and music critics, I started working with Max, confident and full of energy—thrilled, too, when he accepted all my recommendations. The public got to see what we had come up with in early December 1959, when the *New York Times* carried a flashy front-page feature titled "Final Design," complete with architect's rendering of the hall.

But soon after that, things quickly unraveled. The public cried, "Elitism!" how "unconscionable," the newspapers ranted, to provide fewer seats than in

Carnegie Hall, the Philharmonic's current home, then scheduled for demolition. Without consulting me, the beleaguered building committee told Max that 2,400 seats (the maximum my surveys showed possible) wouldn't do—he must come up with at least 2,600, still fewer than Carnegie's 2,746 seats but perhaps enough to quell public outrage (he eventually crammed in 2,646). To achieve extra seating without having to redesign the whole hall, Max nudged me into agreeing to bow the side balconies, but failed to tell me of a change in their orientation—they would now slope steeply downward rather than extend horizontally, on the level. Nor could he come up with a way to include a limited number of acoustical panels above the stage and over the first few rows of seats, a much-praised enhancement that I had recently designed for installation in the Tanglewood Music Shed, summer home of the Boston Symphony Orchestra. A few months before the grand opening, he called for the panels to extend over the entire seating area. I acquiesced once again, but only after he assured me the panels would be installed in such a way that all or part of them could be pulled up to the ceiling if the scheme proved acoustically unworkable. Then, after a bad dream in which he saw swaying panels collide during an earthquake, Max instructed the builder to weld them together into a gigantic raft that could be moved neither up nor down. Finally, the building committee, in a last-ditch effort to reduce the already sky-high cost overrun, rejected our plan to add the kinds of surface irregularities to side walls and ceiling (niches, statues, coffers, and the like) that are acoustically crucial to the best concert halls, and hired an interior decorator to cover up the lack of adornment.

Small wonder the butterflies in my stomach are starting to bite like gnats. I've done my best, but the people around me seem completely unaware of the delicate balances that can turn an acoustical gem into a rhinestone and vice versa. Yet they all have the highest of expectations. And I, too, feel hopeful that, though the hall won't be perfect tonight, any problems can be fixed in the months ahead.

The trial concert the night before has gone well, with an audience consisting of workers from our construction crews, Lincoln Center members, the architects, and some music critics—all appreciative, yet biding their time till opening night, when the performance will be sure to set off a groundswell of either enthusiasm or hostility; indifference is simply not in the cards. Our first hint of trouble comes when the Philharmonic's music director, Leonard Bernstein, remarks that he "would like to have the panels over the stage higher." The audience files in for the grand opening—some pausing to chat and sip champagne, others scurrying to their seats, most thrilled to be in the inauguration of Lincoln Center, all eagerly awaiting Bernstein's familiar stride to the podium.

The program features excerpts from Beethoven's *Missa Solemnis* and Mahler's Eighth Symphony, a double-sized platform, an oversized orchestra, three choruses, and a dozen or more vocal soloists. During musical climaxes and at forte levels, the chorus voices come through distorted and out of balance. Intermission talk revolves around sound quality; adjectives range from "magnificent" to "terrible," and everything in between. Some well-known conductors and composers chime in. Leopold Stokowski pronounces the sound outstanding in every way; Virgil Thomson senses something wrong with the lower ranges. Roy Harris worries about excessive highs; Julius Rudel and Max Rudolf point to occasional mushiness, lack of clarity in choral tone, and a perceptible echo. A few like Samuel Barber urge caution, saying it's too early to judge.

For me, the evening can't end soon enough. I head back to my hotel with a splitting headache triggered by the blare of the orchestra and the choruses and that spot in the Mahler where a percussionist strikes a rail with a sledgehammer.

In the aftermath, "debacle," "disaster," and "catastrophe" become labels of choice. Acoustics—a science often misunderstood and generally taken for granted—provides grist for cocktail chatter: everyone is now acoustics-struck; no one is without an opinion. People talk less about how Bernstein conducts or Cliburn plays than about how well (or not) the bass projects and how today's sound compares with yesterday's. Public reaction follows the contours of a fever chart—now up, now down. Harold Schonberg, principal music critic for the *New York Times,* injects a note of optimism. "I have grown to like the sound in Philharmonic Hall," he writes a few months after the opening. "When its central problem [weakness in the bass] is overcome, Philharmonic Hall will be one of the world's great auditoriums. [I]t will be a hall of unusual clarity and honesty in which musical strands are never obscured and in which sound has remarkable presence."

My colleagues and I get right to work measuring, analyzing spaces and contours, proposing adjustments, and prescribing ways to improve the bass projection. Lincoln Center management is receptive at first, but their president, William Schuman, wracked by doubt and desperate that there be no more foul-ups, asks the architect to assemble a committee of acousticians to evaluate the hall and make alternate recommendations. Schonberg changes his favorable opinion of the acoustics when a member of this committee declares that he has come on board essentially to clean up the mess I have left behind—and in the process, he adds, "to save the acoustics profession."

There's enough blame to go around, of course, but by now I've become a convenient scapegoat. My dream of a great hall and my reputation as an acoustician both appear to be going up in smoke. I square my shoulders and press ahead with other work regardless. Nearly four decades will pass before the *New York Times* comments that "the success of the new Tokyo concert halls can be seen as a vindication for Dr. Leo L. Beranek, who received years of negative publicity after the 1962 opening of New York's Philharmonic Hall." A redemption more than a vindication, perhaps, but I'll take it anyway.

Why begin my memoir with a tale of colossal failure? Certainly, because it stands out in my memory. But, more important, because I learned much from this failure—in particular, it gave me a chance to pause, to reflect, to sort things out, to regain confidence, and to acquire new perspectives. Although my account of events that follow may appear at times to straddle a line between fact and fiction, understatement and hyperbole, the credible and the unlikely, I've taken pains to lay everything out just as I remember it and as my records show. Friends and associates have helped me winnow out the chaff and train my sights on the most important, most illustrative, and most entertaining.

I hope my story will appeal to a variety of readers, including historians, engineers, and executives. But, even more, I hope that it will strike a chord with lay readers whose experiences may differ from my own, but who can relate to the roller-coaster swings—the successes and failures, joys and sorrows—that life throws our way. I hope that pulling together what worked so well for me—and what clearly did not—will prove useful to them.

Riding the Waves

1 Iowa and Beyond: From Bumblebees to Ivy

It was a typical early fall day in an Iowa township: farmers were bringing produce to market, salesmen were coming in by train, and the townsfolk were making their daily trek to the post office. But, five blocks off Main Street, the Beranek home was all aflutter with excitement, worry, and anticipation. A few days earlier, our local doctor, the only one for miles around, had warned Mother that her baby would be large. The family fretted about possible complications, their anxiety deepened by the death of a neighbor woman a few weeks before following an attempted breech birth. Mother's younger sister came over to help, as did a neighbor with midwifery experience. Nobody out here went to a hospital except in the direst of emergencies; the closest facility was 15 miles away—several hours by horse and carriage over rough dirt roads. Dad was busy that day and asked to be called only if he had to be. The doctor was summoned when Mother's contractions grew faster and stronger, but he arrived too late—to find Leo Leroy Beranek, a bouncing eight-pound baby, already wrapped and raising quite a ruckus.

The date was September 15, 1914; the place Solon, Iowa, a town of about 400 inhabitants. My father, Edward, and his father had run a livery stable that went bust shortly after the arrival of Henry Ford's "Tin Lizzie," the Model T. My mother, Beatrice, had trained at Iowa's State Teacher's College and she taught in nearby elementary schools until she married Dad. In those days, everyone had a nickname. My uncle Stanley was "Strantz" and uncle Fred was "Duke"; a neighbor went by "Saint." My parents hated the other common male names, which were sure to become "Jim," "Bill," and the like, so they chose a name that had no obvious nickname, one they hoped would stick. Leo, meaning "lion," seemed short and sweet enough; to this they added my middle name, Leroy, meaning "king." My last name, Beranek, means "lamb" in Czech. Did the confluence of "lion," "king," and "lamb" presage anything about my life? Without stretching things too far, I think it did.

After the failure of the livery business, Dad was forced to return to the Beranek homestead farm, three miles east of Solon, where he had been born and had lived until age 21. Immigrating from Bohemia in the mid-nineteenth century, my great-grandparents settled on this fertile piece of prairie soil in Johnson County. My grandfather, Fred B. Beranek, was born and raised on the homestead farm, which he held on to even after moving to Solon. The farm was the first of three successive farms to which we moved, like nomads, in a period of three years. Mother prompted the last two moves, and it is almost certain that the hard work involved contributed to her early death. Unaccustomed to farm life, she found it tough going—having to do without electricity, cart water from an outdoor pump, and cook on a woodstove, having to use an outdoor privy and to bathe and wash clothes in galvanized washtubs. Heavyset and strong-boned, she had learned the art of cooking, particularly German staples, from her mother. In an era without refrigeration, she had picked up the best techniques for canning fruits and vegetables and for preserving meats in crocks of lard. I remember her tending her large vegetable garden in the sweltering summer heat. But she never showed any sign of frustration and was always loving and considerate. Only now can I imagine how drained she must have felt when she collapsed into bed each night.

Dad was solid and substantial-looking, with a cheerful voice, sincere expression, and flashing, ready smile. He inspired immediate trust in those around him. He had known farming before his Solon fiasco and went back to it, without complaint, to make family ends meet. His days in the fields would start at the crack of dawn—or earlier in winter—with milking the cows and feeding the pigs, with currying and harnessing the horses to plow, sow, or reap and would not end until after dusk when he again would feed and water the farm animals and bed them down for the night.

Farm life was hard enough, but in 1918, just two years after we arrived on the homestead, it took a turn for the worse. Father's younger brother, Stanley, newly married and with no prospects for work elsewhere, moved in with us. Like Mother, Sylvia—Stanley's wife—had never lived on a farm; but unlike Mother, she had never learned how to cook or garden either. Worn out by the additional workload and responsibility, Mother insisted that Dad find another place to farm. The next March, we moved to an acreage not too far from Tipton, at a reduced rent in exchange for the owner's sharing in our profits on produce and hogs sold.

Iowa, as the song goes, is "Where the Tall Corn Grows." And tall it should be—the great glaciers of the Ice Age dumped some of the richest loam anywhere in the continent on this state, up to 20 feet deep some places. Gently rolling, with manicured fields of corn, oats, and alfalfa, well-kept buildings and silos, the landscape exudes prosperity and efficiency in good times. Before the mid-1920s, Iowa was famous for its rainy-day mud roads. Very few were graveled and only the lone, transcontinental Lincoln highway was paved. This meant that travel by automobile between farm and town was possible only when the roads were dry; in wet weather, we turned to horse-drawn vehicles. Farmwork relied on horses—the tractor came into common use only after 1925.

School, Family, Farm Life, Friendships

My introduction to schooling on the outskirts of Tipton happened on a clear, fall-like day in September 1919. Mother dressed me in short pants—no boy wore long pants before age 14—a warm coat, and a stocking cap. With my dog, Collie, beside me, I posed for a photograph with an oversized lunch pail in one hand and a loose-leaf notebook in the other. Mother walked me to school, nearly a mile along a dirt road. Eight-foot cornstalks, laden with heavy ears, whispered to us on either side, accompanied by the sound of crickets and occasional bumblebees. As we approached an intersection, the little red schoolhouse came into view.

All twelve grades gathered in one wooden-walled, high-ceilinged room with rows of desks of various heights. The teacher's desk stood on an elevated platform, with a blackboard behind it and a pot-bellied stove to one side. I was assigned a desk and handed an illustrated reader, a slate, a pad of writing paper, pencils, and crayons, which I stowed carefully beneath my hinged desktop. I had one classmate, a girl who hardly ever spoke to me. The teacher spent a half hour with us each day, mainly working on our manual skills. At recess, the smaller boys played tag while the older ones batted a baseball around; the girls mingled to one side, skipping rope and playing hopscotch. I learned that I could outrun boys my age, and I used my speed as a defense against bullying. Later on, in high school, I would put this fleetness of foot to even better use as an all-state competitor in the 100-yard dash.

The rest of this first school year remains mostly a blank, except for the changeable weather. The daily back-and-forth trek was brightened by ground squirrels racing across the road, and the darting about of red-winged black-birds and sparrows. Rainy days were miserable: I donned boots, slicker, and a yellow rain hat that couldn't keep the wind-driven rain off my face. I trudged through snowdrifts up to my knees, slogged by fallen cornstalks (now minus their ears), and marveled at an occasional "V" of geese flying overhead. I found some small comfort in overshoes, wool socks, a heavy coat, mufflers, and woolen gloves. If a blizzard raged or the snowdrifts threatened to swallow us whole, the teacher's voice—"School called off for today!"—came across our party phone line.

My second year coincided with some dramatic changes in Iowa's school system, with the closing of most multi-grade schools and a phased transition to the modern "consolidated" school system. The school closest to our farm was in Tipton, about 10 miles distant. It was decided that I would enter first grade, counting the previous year as kindergarten.

Getting to school continued to be an adventure in itself. The roads became quagmires in spring and after each rain. Every morning about 7 AM, a school bus stopped at the end of the driveway, 200 feet or so from our farmhouse. And what a bus it was. A team of two horses pulled what looked like an over-sized long box above a high, four-wheel chassis. Inside, along both sides were benches, each about 14 inches wide with an aisle between. We kids scrambled in through a narrow door at the rear. The driver sat in a separate compartment up front, under a roof with side doors; the reins used to guide his horses ran through a small opening. In the wintertime, a kerosene oil heater stood at the front end of the aisle in our compartment. Little if any ventilation got in. The windows were kept tightly shut, and I imagine it was only on account of cracks here and there that we escaped asphyxiation. The ride was just over 2 hours each way, and we generally arrived at school not long after 9 AM. As we bumped along, we passed time telling stories, "finding the button," and play-wrestling. Many of us brought cookies and thermos bottles full of warm milk. The older students tried to keep order, and I learned a little about democracy and sharing on these rides. On winter afternoons, we always arrived home after dark. Ten or more of us crammed into the bus, and we readily took on each other's colds, measles, and mumps, reducing our number by half on some winter days.

Our Tipton stay did not last long. In mid-November 1920, when Mother became pregnant, she decided that going through childbirth in a remote farmhouse without help would be just too difficult (no one thought of going to a hospital in those days). She insisted we move back to Solon to be near her family, although her sister there, Maime, had a family of her own and could not be expected to help. A peak in demand had raised farmland prices to astronomical levels, but Dad found a farm for sale about a mile west of town; somehow, he managed to take out a mortgage and acquire machinery and farm animals. We moved in March 1921.

On entering third grade in September 1922, I found the assignments so easy that my teachers decided I should proceed straight to fourth grade, where the courses were spelling, reading, writing, arithmetic, geography, and language. Although the move was difficult both on an academic and on a social level—I was quite a bit younger than most of my classmates—I got by all right. Once in fifth grade, we were asked to think up an original story and tell it out loud. After I told mine, the class begged for more and the teacher allowed me to weave yet another yarn for their collective entertainment. At recess, the older boys played baseball or football. I tried to avoid the rib-crushing "pile-ons" where one boy would be tackled and the others would flop on top.

During a game of tag, I got a nickname that stuck with me through the remainder of my years in Solon. We had two boundary lines: one side of the two-story brick schoolhouse and the sidewalk about 30 feet in front. One boy was designated "it" and stood between the limits. The others, usually six to eight of us, would on command from "it" run from one boundary to the other, and "it" would attempt to tag someone; the first to be tagged became "it." I came up with a bit of doggerel—"Chi-Chi Mr. Punchi, / Chi-Chi Mr. Nello, / Come catch me if you can, / I am Mr. Punchinello"—flaunting my relative swiftness, here and in other games. Soon I was called "Punch," a nickname I did not relish, but probably deserved.

Asked to go to Grandmother Beranek's house on the afternoon of August 27, 1921, I discovered later I had a baby brother named "Lyle Edward." I found Mother and Lyle in bed, tended to by Aunt Maime. Mother seemed very happy, and so did Dad. I saw the baby up close and touched his hand.

I loved the farm in summertime—its smells, its bustle, its warmth. Not far from the main house were two barns and two sheds. Beyond that grew the crops, and still farther out lay a large pasture for grazing cattle. A creek flowed through the middle of the pasture. Though its bed was four feet deep, there was rarely more than six inches of water in it, except after a drenching rainstorm when the creek overflowed its banks to form a short-lived river. I loved to run barefoot in the creek, catching minnows, chasing birds that came to bathe, and drink the clear, clean spring water that trickled from cracks in the banks. I watched with delight as the tree squirrels, ground squirrels, ground hogs, and gophers moved about in their distinctive ways—some quickly, some furtively, some deliberately, but all seeming to have a keen sense of their place in the natural world.

The plowed fields tended to dry out in the summer heat, when dust devils would form and spiral to heights of a hundred feet or more, sometimes catching me unawares. I remember wiping dust from my eyes and shaking it from my hair and clothes like some character out of the *Arabian Nights*. One summer day, a thunderstorm moved in suddenly, and before I would run home, a bolt of lightning crashed into a tree some 20 feet away, blinding and deafening me for five minutes or so. On hot days, I often went barefoot and once stepped on a rusty nail. Fearing I would come down with lockjaw, my parents rushed me into town. Doctor Nedilicky cauterized the wound with an acid-dipped Q-tip—eliciting a loud yelp from me.—He then applied a soothing flaxseed poultice held in place by a wide bandage.

The farm animals—boars, bulls, and roosters—gave me my earliest lessons in reproduction. I watched the mating rituals with much curiosity, even awe. My parents warned me not to get anywhere near the boars and bulls; a neighbor man had been knocked down and nearly killed by a boar, and stories circulated about bulls charging their owners. The purebred bull on our farm had a large ring in his nose, and when Dad moved him from one area to another, he used a six-foot pole with a hook looped through the nose ring to keep the animal obedient. One day, it dawned on me that babies came from mothers through contact with fathers, and I shared this epiphany with a neighbor boy.

Farm boys learn to tinker at an early age, and I was no exception. I often wonder how much that experience influenced my decision to become an engineer. My special interest in communications engineering almost certainly began in June 1924, when Dad came home with a Crosby one-vacuum-tube radio receiver set that ran on telephone batteries. Using headphones, three people could listen at the same time. I all but devoured the instructions for assembling the set and getting it to work, and gradually came to grasp how radio waves behaved. I installed the antenna and a ground rod, as well as insulating strips under the window to lead wires into the house. I tuned in to a host of things: national news, weather reports, music, political debates. One station—WOS in Jefferson City, Missouri—came through particularly well. I listened every night after doing my homework. One of the most popular entertainers at the time was a jazz pianist, Harry M. Snodgrass, billed as the "King of the Ivories," whose programs originated from the Missouri State Penitentiary, where he was incarcerated. I was glad when the governor pardoned him, but sorry when his last program aired on January 14, 1925.

Grandmother Anna Beranek was a striking woman. Tall, strongly built, and with prematurely gray hair, she was a center of attention at any gathering. Grandfather Beranek had the same hefty build as my father and sported a distinguished mustache. Widely respected, he served as mayor of Solon for many years, up to the time of his death. Grandmother always said she had married him when she found out what a good kisser he was—this in a high-school play, where they had been cast as starry-eyed lovers. I remember that cheery kitchen of hers, where she lovingly fashioned her much-admired "kolaches"—a Czech pastry usually filled with peaches, cherries, or blackberries, but sometimes with my favorite filling—poppy seeds.

Soon after we moved to Solon, Dad bought me a pony, which I named simply "Pony," and which he fitted with a western saddle. I rode him to and from school, keeping him during the day in a barn behind Grandmother Beranek's, not far from the schoolhouse. A neighbor boy also had a pony and a number of times, especially on Saturdays, we went riding together. I remember we once found an abandoned shed, about 5 miles away. We proceeded to kick and pound it to pieces, leaving a pile of rubble behind, mostly boards and beams. I still wonder how we escaped injury or even death when the structure collapsed around us.

Another day I was less lucky, but lucky enough, as it turned out. Pony suddenly decided to cut under a barbed wire that had been stretched over the top of a row of posts to keep horses out. The wire was just high enough to clear the horn on my saddle. I quickly ducked to one side and escaped with just a cut on my outer arm from shoulder to elbow. If I had not ducked, the wire would have slashed into my stomach and, held in place by the back of the saddle, I almost certainly would have been killed. Sometimes life is a matter of luck—or a quick reaction.

To supplement his farm income, my father learned auctioneering. He drove us over in the Model T Ford and assigned me to place number labels on animals up for auction. It was good fun. I was highly susceptible to motion sickness, however, and avoided merry-go-rounds and even swings. Every trip with Dad, I would get carsick and we would have to stop by the side of the road. My affliction did not clear up until many years later, when I started to ride in airplanes.

Mother also found a way to help meet our staggering mortgage payments. She assembled a flock of purebred Rhode Island Red chickens, from which she collected eggs to sell to merchants in Solon and to people around the state who wanted to develop their own flocks. She advertised in the *Wallace Farmer* that she had purebred "eggs for hatching," which she could ship by post to your doorstep—no need for a long trek over to the farm. I helped her wrap eggs in excelsior and lay them out in neat circular rows in shipping baskets. While managing her egg business, Mother continued to help milk the cows, tend her large vegetable garden, do the laundry, make beds, cook, encourage me to study each evening, and even find some time to play piano. She always worried about what would happen if Dad were to die and she were left on her own to provide for herself and two children. She took a correspondence course in bookkeeping, convinced that this would guarantee her employability if her worst fears materialized. I can remember her hunched over a table late in the evening, entering figures on lined paper and underlining sums with red ink and a ruler.

A Tragedy, Then Winds of Change

Christmas 1925 was a happy time for the Beraneks. The presents under the tree exceeded my expectations and we all—aunts and uncles included—headed over to Grandmother Beranek's on Christmas Eve to enjoy a special dinner, featuring Iowa's traditional oyster stew. Afterward, we attended services at the Methodist-Episcopal church, followed by midnight Mass at the Catholic church. Grandfather Beranek and his three sons were Methodist; Grandmother Beranek and my mother were Catholic. We avoided competing loyalties by acknowledging, if not always strictly observing, the core traditions of each.

Three weeks later, a hard freeze settled in, unusual even for mid-January on the prairies. Winter often held us in its grip, but never quite like this. Our old pot-bellied stove wheezed and seemed ready to give up the ghost. We huddled close for warmth as films of ice crept along the inside of the windowpanes. Mother had come down with a severe cold, and it only grew worse as we struggled to keep the elements at bay. Doctor Nedilicky had been to the house two days before, left medicine, and asked her to stay in bed. Her sister Maime came to tend to household essentials.

Monday afternoon, January 26, Dad was taking me to town to attend to business when someone brought news that drained the blood from his face. He turned to me and said, "Let's go!" His jaw set, he said nothing as we sped home. We reached Mother's bedside just in time. She breathed her last with Dad's reassuring hands on her shoulders. I went over, kissed her forehead, and broke down in tears. The attentive priest, distraught relatives, undertakers shuffling about in hushed, respectful tones—all passed me by in a blur. Yet I didn't feel hopeless or lost. Some weeks before, Mother had taken me aside to say I should study hard, get ready for college, and make a decent life for myself. To this day, I remember where she sat and where I stood to receive this sage advice, which I took very much to heart.

Two months after Mother's death, Dad sold the farm at auction, taking a huge loss. He also sold Pony; my dog, Collie, went to live with a relative. Father, my four-year-old brother, Lyle, and I moved in with Grandparents Beranek.

All through this, I was doing quite well in school, ranked near the top of the class. But because Dad's political leanings—toward the Democratic Party—were out of synch in the heart of diehard-Republican Iowa, my class-mates used to taunt me with the jingle: "Fried cats and pickled rats are good enough for the Democrats." I stayed mostly to myself, but found considerable companionship in a cousin, Arlo Bittner, who also happened to be my clos-est competitor academically. Arlo was the son of Grandma Beranek's sister, Blanch, who was younger than Dad, her nephew. Soon after Dad bought me a bicycle in the summer of 1926, Arlo and I were breezing around the coun-tryside. But one night, Arlo left home and did not come back. The next day, they found him wandering aimlessly along a country road. He had suffered a mental breakdown, and could not even recognize his parents. He never recov-ered, even after years of treatment in a mental hospital.

I started junior high school, eighth grade, in the fall of 1926 and yearned to be more independent. I didn't bother either my grandparents or my dad about my needs, though, preferring to see what I could do on my own. When I responded to a Real Silk Company ad for salesmen in the *Saturday Evening Post,* the company assigned me—a mere novice—Solon and a nearby vil-lage for my territory. It sent me a leather-bound sales kit with samples of its entire line of stockings and fabrics for silk lingerie and blouses. For every sale I concluded, the buyer would make a down payment and the order would be shipped COD through the mails. The down payment was my commission. Even my lady teachers bought lingerie from me, with giggles and some embarrassment. I made a modest but regular income, and remained a Real Silk salesman for two years.

Passageways to the Future

The Solon School band leader urged me to learn an instrument and join. I chose drums. Father bought me a marching drum, and the band leader taught me to play. I grew to be reasonably proficient. After a year, Dad purchased a set of trap drums from a retired professional musician, and I continued my lessons on them. I practiced after classes in the basement of the school building. Trap drums would later help me earn my way through college.

My freshman year in high school was my last in Solon. Dad married a woman from a neighboring village—Frances—and we moved to Mount Vernon, some 12 miles away, where he became co-owner, along with his cousin Gilbert, of Beranek Hardware. Keenly interested in my future, Dad came up with the idea that I should learn how radios work so that I could make some money installing and servicing sets sold in the store. He enrolled me in a radio course offered by the International Correspondence Schools. I took this quite seriously, even building my own radio set. The next year, he arranged for me to work as an unpaid apprentice to the store's serviceman, Francis Pratt, a senior in Cornell College just down the road (founded 12 years before the more famous Cornell in Ithaca, New York). Francis was an opera buff and, as we fixed radios, we played records on a wind-up phonograph. My apprenticeship completed, when Francis moved on, I was able to set up a radio repair shop of my own over Beranek Hardware. I bought a Model T Ford for $50 and soon became known as Mount Vernon's "radioman." Meanwhile I had not forgotten Mother's insistence on a college education. Dad made it clear that, because he was still deeply in debt following the sale of the Solon farm, I must save money for college. My radio repair business was no longer just a pleasant hobby; it was now a means to an end.

Churchgoing was always a part of my early life. In Solon, I went to the Catholic church because both Mother and Grandmother Beranek were members. I enjoyed the pageantry, sometimes sang in the choir, and attended Sunday school. In Mount Vernon, I went to the Methodist-Episcopal church, primarily because Dad and my stepmother were members. I thought of church as a way to meet other young people and to learn more about life and biblical times. I often showed up at church affairs. One evening, a pinochle contest was held in the church basement. There were 48 of us in the room, seated 4 to a table around 12 tables. A wonderful game for large groups, pinochle took two standard decks of cards, combined, for each game. When one game ended, the players moved on to different tables. The rules were complicated and I had never played before, but I got some quick instruction and a few pointers as each game progressed. Unbelievably good hands came my way all evening—beginner's luck, no doubt—and about 10 o'clock, to everybody's amazement, I ended up the winner. My prize was a large turkey, which I donated to a poor family in Mount Vernon.

Starting in my junior year of high school, and for three years thereafter, I played trap drums in a ragtag dance band. My talent—or, rather, rhythmic instinct helped along by music lessons—was spotted by Wilbur Powers, a local electrician known to all as "Polly." He put together a dance combo, "Polly and His Parrots." A man of 40, a little on the heavy side, he was a stupendous saxophone player. It was whispered that, though married, he was always on the make for younger women. Our somewhat mismatched band of six played weekly at dances over at the Moose Lodge in Cedar Rapids. The bass horn player, Jake, was a model of respectability, stationmaster for the Chicago North Western Railway in Mount Vernon. The banjo player, Frank, loved to regale us with stories of his trysts with local married women. The trumpeter, Bob, confined his tall tales to fishing and hunting with dramatic accounts of narrow escapes from wolves and mountain lions. The piano player, Hildred, a thin, modest woman of 30, tried to distance herself from the seamier escapades of her cohorts—and certainly offered no comfort to Polly and Frank, our resident rakes. Busy with schoolwork and radio fixing, I simply had no time for sharing racy stories. The band came to an untimely demise when the rumor surfaced that Polly had gotten a high school senior pregnant. The rumor grew larger than life, as gossip so often does in small towns, and Polly and his wife, Rae, had to leave Mount Vernon. They moved to Wilmington, Delaware, where I visited them on my first trip East.

In school, my competitive streak was starting to show. I forget why, but I signed up for a typewriting class. The girls were honing their skills for office work and there I was, the only boy, intruding on *their* territory. Though willing to put up with me in general terms, what they couldn't abide was my being top performer. I looked forward to rattling through our weekly typing tests at breakneck speed and with high accuracy, leaving the others in my wake—helped along by a set of fingers well limbered through long hours of dance-band drumming. The outcome was as much fun as the tests—"If looks could kill," as they say. I had never been one to worry about popularity; my goal was simply to excel in whatever I took on.

During my senior year, I applied for admission to Cornell College (Mount Vernon) and was accepted. Living at home I had managed to save about $500. In 1931, in the wake of the Stock Market Crash of 1929, panicked runs on banks—and the resulting bank failures—were still commonplace. I recognized that we were on the verge of a deep economic depression, and I worried about the fate of my tuition money on deposit in a local bank. In mid-August, I went to the bank to withdraw $400. The clerk called an officer, who asked me what I wanted to do with it. When I told him I was going to pay a year's tuition at Cornell College, he replied: "If you wanted it for any other purpose, I wouldn't give it to you." I headed directly over to Cornell's financial office and put my money down—in the nick of time. The bank closed its doors permanently the very next day, and all depositors lost their savings. I never saw my remaining $100.

Exercise in Self-Sufficiency

In the middle of my freshman year at Cornell, Dad told me that, because of dwindling business, he had sold his share in Beranek Hardware to his cousin and would move to Cedar Rapids in early March. Now I was *really* on my own—no more free room and board. I lucked out, however, in finding a cheap place to stay for the rest of the school year: Ma Miller's student rooming house, where I lodged at a discount. I applied for and received a scholarship for sophomore year, although I still had to come up with $60 per semester for tuition.

I worked as a hired hand on a small farm for two summers, 1932 and 1933, not only to earn my keep but also to better my health and physical stamina. The farm lay to the south of Mount Vernon and my duties there fully tested the limits of my strength. To kill weeds around rows of corn, I walked from 8 AM to 5 PM behind a horse-drawn cultivator, moving slowly but deliberately over ground that the plow had just stirred up in temperatures exceeding 100 degrees on some days. Even though the noon meal meant an hour off, by evening, having rubbed all day against denim overalls, my sweaty legs developed painful chafes, which I salved from a can labeled "For man or beast." Corn cultivation stopped about July 4, but then the oats had to be harvested. I followed behind a binder, operated by my boss, each day plunking down hundreds of bundles (sheaves), six to a shock. Next, hay had to be cut and, after drying, stacked high with a pitchfork onto a wagon and hauled to a barn for transfer into the haymow. Gardens and melon patches had to be weeded, animals watered and fed. On some days, fences had to be repaired and rings put in the noses of hogs to discourage them from rooting. In late July, we harvested melons and picked berries to take to market. Yet, a few evenings each week, I still found the energy to jump into my Model T Ford and head over to Mount Vernon to fix radios or play in the town band.

Because we helped each other out at threshing time, I got to know most of the farmers in that part of the country. Although Prohibition was in full swing, when one of our neighbors took up bootlegging, we turned a blind eye. I dated a neighbor's daughter, who had been in high school with me, and on Sundays, often visited my grandparents in Solon. But I seldom got over to Cedar Rapids to visit Dad and my stepmother. The 20-plus miles seemed like an awfully long way in those days.

I ended the summers tanned and far stronger than I started out. Once, I even hefted a 160-pound keg of nails. I sometimes wonder how much those summers contributed to making my life as free as it has been from illness, and as active in my advancing years—with most of my joints and "marbles" still intact.

The college-owned dormitories and dining halls were way beyond my means. So, when Ma Miller offered no further discounts, I had to find another place to live. In late August, right before the start of sophomore year, I learned that three other students—seniors—had made arrangements to live in two large, unfurnished, unheated rooms over a bakery on Main Street. They invited me to join them. We needed furniture, a stove, cookware, and dishes. Freddie Katz came up with most of the furniture—four beds, four bureaus, four desks, and a half dozen chairs—on loan from his father's secondhand furniture store in Cedar Rapids. I borrowed an oil-burning stove from Beranek Hardware. Other items came from our parents' homes or were borrowed from friends. We each paid $4.50 a month in rent, and put in an extra buck or two for breakfasts and evening meals. We cooked one warm course each evening on the stove's single burner. I arranged for the bakery downstairs to pass along their one-day-old bakery goods for a dollar a week. Wilbur Smith dated a college woman who lived on campus and worked in a dormitory kitchen there. From time to time, she filched a whole roast chicken and passed it through a window to Wilbur. Leo Phearman's farm family sent eggs, smoked ham, and fresh fruit from their orchard. Freddie often brought packaged food to the table. How he acquired this we never knew, but I suspect his father helped out. On occasion, I dipped into my meager savings to bring in extra chow. I earned my noon meals by waiting tables in the Fair Deal, a restaurant just down the street.

As a sophomore, I was invited to be a member of Mort Glosser's college dance band. We played at campus dances on Saturday nights. Although my income from this and from my radio repair business took care of the year's expenses, I saw problems ahead. The repair business had fallen off—the Great Depression was deepening—and Cornell's scholarship stipend was smaller than it had been the year before. Because I could see no way to earn what I needed, and because no student in those days—in Iowa, at least—ever thought about borrowing, I resigned myself to enrolling in just one class the next year, mathematics.

The Outside World Beckons

Invited to go with the family of the local dentist, Lou Bigger, to Chicago's world fair in August 1933, I jumped at the chance. The Century of Progress opened my eyes to the world beyond Iowa. I wandered from one exhibit to another, almost in a trance. The ones I remember the most showed manufacturing, such as brand-new tires all wrapped in paper. The stunningly illuminated Electrical Building showed electricity being generated and distributed, with a fireworks display every night. "The introduction of electricity in our daily life is the greatest factor in human progress," announced General Electric, presenting a dramatic set of murals to illustrate its claim. Ford, Chrysler, and General Motors exhibited their latest cars and showed motion pictures of the assembly process. Pabst Blue Ribbon, Schlitz, Budweiser, and Old Heidelberg set up their wares in huge tents filled with tables. Each tent had a stage at one end, where a popular "big band" of the day kept visitors tapping their toes.

The best surprise was the opportunity to hear, not one, but four hour-long outdoor symphony concerts each day. The Chicago Symphony, sponsored by Swift, performed twice a day, as did the Detroit Symphony, sponsored by Ford. I made it a point to attend a concert every day.

The foreign pavilions gave a panoramic sweep of world cultures enthralling to a sheltered Iowa youngster brought up among German-American farmers. I marveled at the Italian, Chinese, Japanese, Ukrainian, Moroccan, Belgian, and French exhibits. High in the air, the Sky Ride Car, a double-decker gondola, sailed from one end of the fairgrounds to the other across a lagoon. Because I had so little money, I lived on popcorn, hamburgers, and two beers a day. When I got home, I calculated my total expenses for four days at $12.00. I went again the following summer.

In the fall of 1933, learning that Albert's, a dry-cleaning and laundry business in town, would house a student willing to help out, I applied and was accepted. I slept in a back room with bags of dry-cleaned clothes hanging some two feet above my bed. I was expected to start the steam boiler and sweep the floors each morning before the owners arrived.

College High Jinks

For Midwesterners, Halloween was, and probably still is, a major event. I can remember when I was living with Grandmother Beranek in Solon how high the excitement ran, how pranks would sometimes morph into vandalism, and how the townsfolk seemed resigned to this as part of an age-old custom. Main Street usually looked bombed out the morning after. Soap smears covered store windows, wheels from parked cars ended up on rooftops, and sidewalk benches were scattered everywhere. The most adventurous pranksters had absconded with outdoor privies from private homes (there was no town sewer system in 1926), lining them up in not-so-neat rows down the center of Main Street.

Things were pretty much the same in Mount Vernon, although it did have a sewer system. On Halloween in 1933, about 10 PM, I heard a knock on the front door of Albert's just as I was turning in. Six Cornell students stood in the doorway and asked me to join them in commandeering a large privy behind a home on the outskirts of town and moving it to the college dean's front porch. Dean Albion King was an officious sort, with no friends among the students. When I asked my colleagues in crime how they planned to transport the privy, they said they would simply carry it. Slipping into engineering mode, I quickly calculated the weight and reported the disappointing news: it was far too heavy to be carried. Then I got an idea. Behind the local telephone building, not far from Albert's, was a four-wheel flatbed trailer, which the telephone company used to transport telephone poles. I proposed that we "borrow" it, which we did, and we stealthily headed out to fetch our loot.

The privy was even larger and heavier than I'd calculated, but somehow we managed to tilt it onto the trailer and push and pull it to campus. Within a few hundred yards of the dean's house, a police officer I knew stopped us. I had repaired his radio once. He asked where we thought we were going with that privy. To Dean King's front porch, I told him, knowing that King's unpopularity extended well beyond the student body. "If you put it anywhere else," the officer replied, "I will arrest you." After a struggle, we managed to stand the privy on the dean's wide, covered porch without waking anyone up—and to roll the trailer back to where we found it with no one the wiser.

The next morning, Dean King came to school fuming, exactly as we'd hoped. His suspicion fell right away on the likely perpetrators—the Deltas, a fraternity made up mostly of athletes. He grilled them, one by one, in his office. But none had been in our group, and, in the end, the dean failed to identify any of the culprits. He never suspected that I, of all people, had been willing not only to embark on such a disreputable scheme but also to make sure that it succeeded.

Radio Electronics—A Hobby Becomes a Vocation

I soon became close friends with one of my mathematics classmates, Harold Ericson, a tall, slender fellow with a pleasant manner and disposition whose father owned a telephone company in Hector, Minnesota. Harold was a ham (licensed amateur radio station operator) and knew lots about radio receivers and transmitters. He urged me to get an amateur license, too, so that I could share in the use of his transmitter. This meant I had to learn Morse code. Harold loaned me a small code-sounding machine, and I learned the dits and dahs (dots and dashes) of the Morse alphabet. When I felt confident enough to pass the test, I went by bus to a Federal Communications Commission (FCC) examination center in Des Moines. All candidates had to show they could send and receive code at not less than 10 words a minute. I squeaked by, earning the call letters "WRER," which I could use as my signature anytime I broadcast. This experience came in handy later when I went to graduate school.

At some point in the fall of 1933, the head of Cornell's speech department happened to mention how helpful it would be for his students—as a measure of their progress—to record their voices before and after their year of speech training. I later chanced on an advertisement in a radio magazine for a small recording machine, which, I found out, could produce embossed aluminum records at a cost of about 50 cents apiece. I went back to the professor and told him that I would buy the machine out of my own pocket if he would commission me to produce a 5-inch aluminum record for each student at a dollar apiece. He agreed and, in the small studio I built in one corner of a physics lab, I recorded every member of his class before and after a semester of speech training. This brought me in contact with the field of acoustics for the first time. I learned something about acoustics from a book in the college library, little realizing that it would become a major part of my professional life in the years ahead.

As the spring of 1934 approached, I found myself in financial straits once again and realized I would need to find a full-time job to accumulate some savings. In April, I applied to the fledgling Collins Radio Company in Cedar Rapids. The young president, Arthur Collins, interviewed me personally and offered me a position at 14 dollars a week, starting May 15. Arthur was the son of a wealthy real estate dealer, M. H. Collins, who had managed to avoid the worst of the bank and stock failures during the Depression. Arthur was known locally as boy inventor and radio wizard extraordinaire. Arthur's father set him up in Collins Radio, which at first manufactured only transmitters for amateurs. His breakthrough came in 1928, when he was chosen to equip Admiral Richard Byrd's expedition to the South Pole with a small transmitter that ended up performing magnificently. I became a friend of Arthur's and he invited me to his home several times to look over his amateur radio equipment and to share an evening with his family.

Now that I was leaving Mount Vernon, I decided to sell my radio repair business, which I did, to Harold Ericson for $40. I stored my drums in Bigger's attic and headed for Cedar Rapids, where I rented a room for a dollar a week in the home of an elderly couple. I remember keeping close tabs on the clock, as we had warm water available just two hours a day.

Collins Radio was moving ahead with some fairly adventurous marketing strategies at the time, particularly with a view toward broadening its clientele for the manufacture and sale of sound systems. One possibility was that funeral homes might play recorded music during services. I was assigned to go on sales expeditions with an old salesman, Jim Thompson, who was about 35, and had little in the way of technical know-how. My job was to demonstrate equipment, answer technical questions, and help plan the kind of system each client needed. Collins gave us an old Cadillac, a four-door sedan, to travel around in. We visited funeral homes all across Iowa, staying in low-cost motels as we roamed the countryside.

When hot weather approached, Jim always wanted to find the nearest public beach or swimming hole. He also had quite an eye for the ladies. After making several sales pitches in a town, he would take me out "cruising." If he glimpsed a pair of presentable young women walking down the street, he would pull the Cadillac over and strike up a conversation. On one such occasion, the women joined us for dinner, and because it was a very hot day, Jim asked them where we could all go for a swim afterward. As dinner progressed and the beer and talk flowed more freely, he proposed that we invite some of their friends along and head over to the nearest pond for an evening of skinny-dipping. The perpetual charmer, Jim put together a willing group of six or so. When I declined to join in, he dropped me off at our motel—whether he was miffed with me for party pooping or happy to have the ladies all to himself, it was hard to tell. The next day, without going into specifics, he boasted about what a great time they had had. Partly as a result of such distractions, our sales expedition was unprofitable and Collins fired Jim in September.

The company asked me to stay on as an assistant in the engineering department. Three new engineers arrived at Collins at the same time—Frank Davis, Merrill Smith, and Roger Pieracci—all with master's degrees in engineering from prominent universities. When I met them on their tour of the plant, the subject of housing came up, and the four of us decided to join forces and look for someplace to live near work. We found a very nice second-floor apartment that had two bedrooms, each with twin beds, a living room, a kitchen, and a wide hallways in a house on the nearby streetcar line—and all for $40 a month. We shared the household chores, and my diary shows that, about once a month, my turn came to scrub the floors and vacuum the rugs.

I kept very busy outside of work, reading books that ranged from dime mysteries to engineering texts, and going out on dates. On one such date, I met Florence (Floss) Martin, a beautiful, slim woman some three years younger than I. Floss was attending business school in Cedar Rapids and lived with her aunt in a modest second-floor apartment not far from me. She and I hit it off from the start, finding plenty to talk about, and I began seeing her regularly.

We liked going to movies and kicking up our heels at Danceland, the city ballroom. By the time I returned to Cornell, Floss was my steady girlfriend, and we had even talked about getting married someday.

That fall in Cedar Rapids, I took German lessons because I had barely scraped by with a C in German at Cornell. My prospective teacher was a recent immigrant from Germany. Mr. Merner said he would teach me as long as I was willing to prepare each lesson fully and to show up at his place twice a week, promptly at 7 PM, for an hour and a half. I agreed. He began by pulling a book of German fables off his bookshelf. I studied the text so thoroughly I came to know every punctuation mark. (Quite by chance at the end of my next semester at Cornell, the professor chose to read to the class one of those very fables. To his amazement, I handed in a perfect translation.)

On New Year's Eve 1934, after working all day at Collins Radio, I rushed over to pick up Floss for a night of celebrating at Danceland. At midnight, we shouted, blew horns, and kissed, then went with my roommate Roger and his girlfriend, Bernice, to the Play-Mor Hall, where we danced to Al Morey's band until 1:20 AM. After that, we headed over to the Iowa Theater to watch the end of a movie, followed by a stop at the Montrose Hotel to enjoy yet another dance band. The evening—or morning, by this time—ended at the Butterfly Sweet Shop for coffee and cake. I dropped Floss home and fell into my own bed at 5:10 AM. My notes say: "What a grand ending to the old year. Finest year ever for me—God grant me more. What a great thing life is!"

In January, I made plans to return to Cornell. Arthur Collins appeared sorry to see me go, but I had put aside what I had aimed to—a pot of savings to help me finish up college—and along the way I had learned a lot and gotten to know some interesting people. I arranged with Harold Ericson to share in (now) his radio repair business back in Mount Vernon. Cornell awarded me a second-semester scholarship of $112.50, which meant that I only had to find $87.50 for tuition. Another lucky break: fellow student Richard Rhode asked me to join his popular college dance band. I played drums with them about once a week, earning $4.00 each time.

A Momentous Encounter

At the end of the semester, I went back to Cedar Rapids to spend the summer working once again for Collins Radio. I moved into the same dollar-a-week room that I had previously occupied and, always a planner, I started thinking about what I would do after graduation, just over a year away, when something happened that led me in a direction I never expected. On Friday evening, August 16, 1935, I drove to Mount Vernon on the Lincoln Highway, which then went from New York to San Francisco, passing through Mount Vernon along its Main Street. I spent the night on the back-porch swing of the Bigger family's home. After lunch, having passed the morning at the Cornell Library reading technical periodicals, I was strolling along Main Street when I came across a Cadillac with Massachusetts plates standing at the curb with a flat tire. Beside it was a well-dressed man looking glum. When I asked him if I could help he jumped at the offer. As I worked away with the jack and lug nuts, we engaged in a friendly exchange. I told him that I was between my junior and senior years at nearby Cornell College and how I wanted to go to graduate school, but could not afford to unless I were to obtain a scholarship. He asked me about my majors and my grades. I cheerfully answered and said that I was planning to submit scholarship requests to the University of Iowa and to the universities in the states surrounding Iowa.

At the mention of my work as a radio repairman, he perked right up. "Radio is my business," he said. He asked for my name, and after responding, I asked for his. "You are Glenn Browning?" I blurted out. "I just read one of your papers on the Browning Tuner in *Radio News* this morning in the library." Suddenly I had a new friend. He wanted to know if I had considered going to Harvard University. "No," I said, and then—before I could catch myself—"that's a rich man's school." Smiling, he informed me that Harvard had more scholarship money to offer than any of the schools I had named. He opened the door to the front seat of the car and took out a pad of paper on which he jotted down the names and addresses of two people at Harvard—one for admissions and the other for scholarships. "When you submit the paperwork," he said, "use me as one of your references." I would learn later that he had spent three years as an instructor at Harvard's engineering school before opening a successful radio manufacturing business in suburban Winchester.

That fall, I sent scholarship applications to various state universities. I also sent one to Harvard. When I wrote to Browning in February 1936, thanking him for letting me use him as a reference, he wrote back that he had already been contacted by Harvard's dean of engineering, had put in a good word for me, and wished me luck.

Letters from the state universities started arriving in March—all of them saying, in effect, that my grade record and references were satisfactory, but that, because there were so many applicants, they could not offer me a scholarship. Then a letter came from Harvard. I opened it slowly, anticipating yet another letdown. Dated March 27, 1936, "I am very happy to tell you that you have been awarded a Gordon McKay Scholarship of four hundred dollars, covering your tuition for study in the Graduate School of Engineering at Harvard University during the academic year 1936–37." When the news got out, I could barely contain my joy. I became an instant celebrity.

Wrapping Up at Cornell

In my senior year, I bought back the radio repair business from Harold Ericson and took up residence at the Neff Funeral Home on Main Street in Mount Vernon, where, in lieu of rent, I helped undertaker William Neff pick up corpses, usually in the middle of the night. To finish up at Cornell that summer, I piled on the subjects: sociology, mathematics, philosophy, German, physics, and art.

At the beginning of the school year, I was elected to the Alpha-Theta-Alpha fraternity, with headquarters in Ma Miller's rooming house just off campus. As part of our hazing, about a dozen of us pledges were given a list of things to get done in one evening: steal some watermelons, pilfer a pig, swipe a girl's panties from a dormitory, and answer a set of tricky questions correctly. The penalty for failure was a dozen or so whacks on the rear with a large paddle—a fate we wanted to avoid at all costs. We drove over to see the farmer I had worked for who raised watermelons and who let us "steal" enough melons to treat our whole fraternity. At one of the girls' dormitories, we stood outside and yelled up our plea for a pair of panties; to our surprise, not one, but two pairs came flying through an upper window. Another farmer I had come to know during my summer stints agreed to let us "pilfer" a pig, on condition that if the animal were not returned, we would pay him $20. With our booty in hand, we got back to the house about two hours after we started out. After correctly answering the tricky questions, we were showered with praise. But the pig got away from us and we couldn't find it in the dark, so the next day we had to fork over $20.

To make extra money in my senior year, I added retail sales to my radio repair business, with a shop on the second floor of a building on Main Street. I convinced the RCA and Atwater Kent radio suppliers to ship me a dozen sets. My shingle over the entryway downstairs read: "Leo L. Beranek, Radios and Service." I found an unemployed man—about 40 years old, intelligent, nicely dressed, and clean looking—to run the sales room whenever I was doing other things, mostly schoolwork. The radios sold well and my repair business picked up, too. I also hired an electrician to help and wired some private homes.

I attended classes during the day and studied in the library in the morning and early afternoons, where I could be certain to avoid interruption. In the late afternoons, I tended to the radio business. That winter, Cornell President Herbert Burgstahler sent out a questionnaire to all students asking how many hours a week we spent on nonacademic activities. The answers were anonymous, and I reported 40 hours. In our compulsory chapel service, Burgstahler announced the results and made a special point of observing that whoever reported spending 40 hours a week outside could not be getting much out of college.

The New Year brought much change. When radio sales dropped off but the house wiring side of my business grew—because my rates were cheaper than those of the Iowa Electric Company and I benefited from subsidies under the Federal Rural Electrification Act—I decided to concentrate on wiring and stop selling radios. An advertisement I placed in the Mount Vernon paper read: "Radio Clean-Up Sale! Friday, Saturday and Monday, January 24–25–27." And, sure enough, I pretty much cleaned out my stock that weekend.

One day every weekend, I would drive over to Cedar Rapids to see Floss. Some Sundays we went to her church, First Christian, and took part in young people's fellowship activities there. Sometimes she would catch the bus to Mount Vernon and we would attend a college social together. We also exchanged letters weekly. Then came the day we went on a picnic with a small group of her relatives. This being the first time I'd met the greater family, I wanted to make a good impression. No such luck. Floss's aunt asked me to drive her car. But, as I pulled out of her parking space, I pressed down too hard on the accelerator, clipping and bending the bumper of the car in front and causing the bumper on our car to fall off altogether. The usual awkward exchange of information and documents was made even more awkward because I had no license to exchange and could only stand there, looking foolish. When the aunt's car went into the garage a few days later, fortunately, nobody asked me to help out with the repair bills.

With graduation not far off, three opportunities for wiring jobs came up at Cornell. In June, I wired the dining-hall addition to Bowman Hall and, a month later, completed the rewiring of Rood House. My third Cornell job was the most ambitious of all. I had convinced the building committee that a central antenna system should be installed in Merner Hall, a new men's dormitory then under construction—at a fixed price of $556.58. Later that summer, the *Cedar Rapids Gazette* announced: "Individual radio outlets [at Merner Hall] are connected with a master antenna system designed and installed by Leo Beranek of Cedar Rapids. The antenna on the roof of the hall is connected with room outlets by a continuous system of wiring in conduits." Each of my early business ventures proved excellent preparation for managing a wartime research laboratory at Harvard, four years later.

I received my bachelor's degree in the summer of 1936, just as I had hoped. I missed Phi Beta Kappa by a tenth of a point, but Cornell made up for that twenty-six years later by naming me an honorary member. Diploma in hand, I sold my radio business to a repairman in Lisbon, Iowa, for $99 at the end of August. With a mix of excitement and nervous anticipation, I started to gear up for what was looking more and more like a risky plunge into the unknown. I had saved about $450, which—along with my scholarship—I was hoping could be stretched to cover rail fare, room, board, and essentials for a year at Harvard. But I was already wondering, *then* what?

2 Harvard: Shaping the Future

It was August 1936 and I was about to enter a completely unknown world with so little money I felt almost overwhelmed. I didn't even know which trains to take to get from Mount Vernon to Cambridge, Massachusetts. I went to our local Union Pacific Depot and looked at schedules, but they showed only Boston. I knew from my atlas that Cambridge was nearby, so I asked the stationmaster. He wasn't much help. "Go to Boston," he said, "there must be local transportation from there." My mathematics professor, Elmer Moots, gave me the address of a former student who was living in Medford, near Cambridge, and was married to John Barnes, a professor in communication engineering at Tufts University. When I wrote Mrs. Barnes, she replied that Cambridge wasn't far from the railway station and that her husband and she would be "happy for you to stay at our house until you find a place. Give us your arrival information and John will pick you up."

What a relief. But I still had to finish my course work and pass examinations to get my bachelor's degree. And I still had to complete the antenna wiring of Merner Hall, a large job in its own right. I had to plan what I would take with me to Cambridge, pay my final bills, and spend a few days visiting family in Cedar Rapids and my grandparents in Solon. Most important, I had to say good-bye to Floss, who was broken up over my move east.

My trip almost ended on September 9 at her apartment. The trouble began when, cutting a label from the collar of a shirt Floss was wearing, I also cut a quarter inch of skin from my knuckle and began to bleed profusely. Floss called several doctors before finding one who would take care of me right away. He used a metal clamp to close the wound and I returned to Mount Vernon. I spent my last night, September 10, at the Neffs' where I had lived the past year. When I woke up around 10 o'clock that night with a red streak starting up my arm, I panicked. Mrs. Neff said I should go see Doctor Hesburg, whose office was just a few blocks away.

I shuddered when I saw his operating table and glass cabinets filled with ominous-looking instruments, but his gray hair and quiet confidence calmed me. He removed the metal clamp, cleaned my sore, sewed the skin together, and applied zinc oxide ointment, a Band-Aid, and finally a bulky cloth bandage over the whole hand. Then he went upstairs and brought down a bread wrapper, which he pulled over my bandaged hand. "You must keep the outer bandage wet for two days," he said, handing me a quart bottle of diluted disinfectant, "then remove it, and a day later take off the Band-Aid, add more zinc oxide and a new Band-Aid." The sore, he pronounced confidently, ought to heal in a week. In those days, there were no antibiotics and healing could be problematic. Dr. Hesburg's bill was $7.00; my train ticket to Boston had cost $25.78.

The Neffs drove me to the railway station. I checked my baggage—a suitcase and trunk—through to Wilmington, Delaware, because I was spending three days there with Polly and Rae Powers, formerly of Mount Vernon. I boarded the train to Chicago, the first leg of my trip. On the next and longest leg, from Chicago to Washington, D.C., I slept sitting up, leaning into the corner between the seat and the window, and every hour waking up to pour liquid on the bandage. The conductors sympathized and made encouraging remarks, while the train bounced and clanked along so roughly that my head and rump grew sore. I bought a sandwich and milk, twice, from a man pushing a food cart through the main aisle.

As the train chugged eastward and a parade of telephone poles flew past, the scenery changed from cornfields, to expansive pastures with dairy cattle, and finally to truck gardens. Most impressive were the Appalachian Mountains, for I had never seen anything besides the rolling hills of the Midwest.

In Washington, I changed from the Baltimore and Ohio to the Pennsylvania Railroad, hoping my baggage would be transferred. I arrived at Wilmington exhausted, but elated to see Polly and Rae on the platform. I collapsed for a full day at their place, pouring liquid on my hand each time I woke up. Polly and Rae then took me to Atlantic City, where we ended up spending three days. The Steel Pier, a truly magical amusement park on Wilmington's famous Boardwalk, was at its historic best—with diving spectaculars, the world's largest miniature railway, a minstrel troupe, and the Varieties of 1936, featuring Lester Cole, Abbott and Costello, Charlie Boyden, and Jimmy Jones's Steel

Pier Orchestra. The Pier was preparing for the Miss America Pageant and we watched some rehearsals. We saw three movies—William Powell and Carole Lombard in *My Man Godfrey;* Fredric March, Lionel Barrymore, and Warner Baxter in *Road to Glory;* and Jessie Mathews and Robert Young in *Love Again*— all included in the price of a single ticket. I left Wilmington rested, healed, and ready to forge ahead with my big adventure.

Feeling My Way

At South Station in Boston, I walked the length of the long platform, past the hot and hissing steam engine before finding John Barnes. A tall, lanky man, he welcomed me warmly: "Wait a few minutes while I fetch my car. The ride to Medford takes only twenty minutes." A porter loaded the trunk and suitcase into his Chevrolet and off we went. Mrs. Barnes—Mabel—was a delight, peppy and full of information as we ate dinner and chatted about our alma mater, Cornell, and my plans for Harvard. The next day, John took me to the bus to Harvard Square.

I got off at the transportation "pill box" at the very moment Franklin D. Roosevelt and James B. Conant, Harvard's president, drove by in a large convertible with the top down and a motorcycle police escort. Harvard was celebrating its tercentenary that week, and President Roosevelt was in town to speak. The hustle-bustle was hard to deal with, but I managed to wend my way from Harvard Yard to the Cruft Laboratory building, where my classes would take place. Superintendent Matt Carley gave me a short tour and took me to see Professor Chaffee, the laboratory director. Chaffee handed me a schedule of classes and pointed me in the direction of Phillips Brooks House, where I would find information on rooming houses.

I walked into a high-ceilinged, mahogany-lined, well-lit, carpeted room, where two attendants, seated behind tables, were busy helping students. When I stood back, waiting for some indication that I was next, several students barged ahead of me. Finally, looking up, one of the attendants observed: "You poor fellow, you are just too timid." He gave me several rooming-house possibilities, but that was the last time I ever went to Phillips Brooks House. I chose a room on the second floor of a wooden three-story house at 16 Irving Street. The rent was 4 dollars a week, for which I got a desk, chair, bed, and wardrobe along with weekly bed-linen service. A pay telephone hung in the entryway downstairs.

Classes started Tuesday, September 29. Imagine a time when electronics meant only radio, and when television was still in the laboratory and PCs were unheard of. The Cruft Laboratory held a special place, even then, in the history of radio communications. It had been established at the time of Marconi's first transatlantic transmissions, between England and Cape Cod. On its roof were two tremendous towers, originally used for worldwide radio ("wireless" at that time) transmissions. In the basement were remnants of Marconi's innovative equipment—Tesla coils, large condenser blocks, and gasoline-driven generators.

My professors varied widely both in background and in temperament. Considered one of the world's authorities on vacuum tubes, Professor (E. Leon) Chaffee was middle-aged with an athletic build despite his widening waistline. Although with eyes distorted by bottle-thick lenses, his good humor and careful preparation made him a classroom favorite. A multimillionaire, whose fortune derived from inventions such as the Pierce oscillator, which helped improve radio reception, Professor G. W. (George Washington) Pierce was short and rotund, nearly bald, with darting eyes. Constantly smiling, he was full of jokes. When a professor arrived at work one morning with a bandage on his neck, Pierce greeted him with "A washed spot never boils." He enjoyed lounging in a big leather chair in his office, musing over world affairs while puffing on an ever-present Cuban stogie. At the other extreme, Professor Harry Mimno was cold, efficient, apparently emotionless, and almost unapproachable; Professor J. Gibbs was average in every way, nondescript and uninspiring.

I discovered that three meals a day at the cafeterias in Harvard Square cost about a dollar, which brought me to the painful realization that the money I'd saved up would not cover my expenses for the year. By mid-October, in order to stretch my meager funds, I cut down to two meals a day, at 50 cents total. My twice-daily trek took me either to the Red Apple Restaurant next to the Harvard Cooperative Society or to the Georgian Restaurant on Dunster Street, two blocks away. In either place, I would go to a large counter to place an order, and, when it was ready, retreat with a tray to one of the granite-topped tables. At breakfast, 15 cents bought me a poached egg, a small glass of orange juice, a piece of toast, and a cup of coffee; at dinner, the blue plate special consisted of meat, potato, and a vegetable, along with a glass of milk and a small piece of cake or pie. I lived that way for the rest of the school year, losing twenty pounds in the process.

I remember the walk from Irving Street to Harvard Square, under the hot sun, in pouring rain, through blinding snow, sleet, and wind, and over sidewalks that could be sizzling or icy, depending on the season. Classes took place in the mornings. I would climb the steps from the entryway to the Cruft lecture room, which seated about 80 students. Up front stood a large table and two immense blackboards that could be electrically elevated one above the other. Nearly everything was written on the boards by the instructor; projections were uncommon. Most days at noon, I would return to my room to study, although some afternoons I stayed for laboratory sessions. The occasional movie cost 20 cents; the subway to Boston 10 cents. When I went to the infirmary to have my teeth checked, the dentist told me that my previous dental care had been inferior and recommended that the dozen fillings in my teeth be replaced.

Floss and I wrote to each other several times a week, in letters filled with talk of love and marriage. Dad and I exchanged letters twice a month. I also wrote Professors Moots and Nelson at Cornell, Grandmother Beranek, and my friends in Mount Vernon.

My closest pals at Irving Street were Philip Goodheim, from upstate New York, and Robert Hull, from Akron, Ohio. Phil, a bookish person, was studying to be a lawyer and Bob, more relaxed, hoped to become a professor of Latin. We would talk politics—a subject almost foreign to me in Iowa; we often shared meals and went to movies together. Bob could name almost any symphony and its composer after listening to just a few seconds of a piece. Every afternoon at about 4 PM, he would play a Beethoven symphony. The scratchy sounds of his gramophone would penetrate just about every wall in the house. Frequently, it was the Ninth, and I came to love it. At the Cruft Laboratory, one of my best friends was Robert (Bob) Watson, from Champaign, Illinois, whose father, Floyd Watson, was a professor of acoustics at the University of Illinois. Later, when I decided that my research work would be in that field, I learned that he was one of the best-known architectural acousticians in the United States.

Added sources of money that year included the sale of my drums—$45, a final payment from Cornell for the Merner Hall antenna wiring job—$143; and cash presents at Christmas—$46. On January 15, 1937, I wrote in my diary: "Have $115 to my name. Have to live to June on it, or find a way to borrow. Hope I get a fellowship [for next year]." I supplemented my funds by borrowing $100 from the Bursar's Office. I also talked to Professor Chaffee about the upcoming year. "We can only provide a scholarship," he said. "We have no money for student assistance."

Proving Myself

I pushed myself to get high grades. The Harvard aura—ancient buildings, tweedy students and professors, the crème de la crème of the academic world—was more than a little intimidating. A Midwestern farm boy, I felt so out of place I wound up working ridiculous hours—with the idea that at least I'd show I could compete with the best of them. I went to every class and studied most other times. During Christmas vacation, I trekked over to the library for periods of in-depth study. Getting ready for Harvard's midyear examinations was a nerve-wracking experience. Because assignments were minimal during Harvard's January "reading period," I dedicated one week to an intensive review of each of my four courses. I approached and undertook the exams with foreboding. Yet, when the grade card came, I had earned three A's and one B. In the second semester, I got three A's and one A+. I was awarded the Master of Science degree in June, but did not go to commencement—my summer job had already started.

The Boston Symphony Orchestra performed in Harvard's Sanders Theatre five Tuesday evenings between October and May. I learned that if I arrived an hour early and stood in line, I could get a student rush ticket for 25 cents. I now was able to hear the greatest classical music in an ideal acoustical setting, an experience that even surpassed what I'd heard when the Chicago Symphony Orchestra performed at Cornell's May Music Festival. I got acquainted with the ticket seller at Sanders; in exchange for a pack of cigarettes each time (cost: 12 cents), he agreed to hold a ticket for me until fifteen minutes before starting time.

Vivid in my memory is George M. Cohan's musical *I'd Rather Be Right,* the first play I attended in Boston (upper balcony: 80 cents), starring Cohan himself. Known as the father of American musical comedy and, before World War I, as "the man who owned Broadway," Cohan had come out of retirement just for this show. A great hit in Boston's pre-Broadway performances, and again in New York and Washington, D.C., it lampooned the Roosevelt Administration, with Cohan playing the president. I especially remember when "Roosevelt" summons his secretary and tells her "Take a law," and when he proposes that the government hire pickpockets to balance the budget because there is nowhere else he can see to add taxes. Cohan could sing and dance like no one I had ever seen or have seen since. A year or so later, the real President Roosevelt bestowed the Congressional Medal of Honor on him, ostensibly for his patriotic song from World War I, *Over There,* but everyone believed it was really for *I'd Rather Be Right.*

During the summer of 1937, I worked at a correspondence training school, the Massachusetts Television Institute—a job arranged by Glenn Browning, the same man who had helped me get a Harvard scholarship. In a building perched directly over railroad tracks on the corner of Boylston Street and Massachusetts Avenue in Boston, a fellow named George (I forget his last name) and I wrote lessons full-time. The unair-conditioned office felt and looked like Hell itself—beastly hot, with billowing plumes of smoke, steam, and soot seeping through the walls and windows. We distracted ourselves from time to time with talk of the latest baseball matchup. Fenway Park was just a few blocks away, and though the Red Sox ended up way out of contention, that season saw the Major League debut of Johnny Doerr and a few other fine players. When we got carried away, our boss would tell us to stop nattering and get back to work. Still, over that two-month period, I wrote six lessons and the examinations to go with each. I hesitate to read one of those documents today, but the experience came in handy when I began to write technical papers and books. My pay for the summer, which ended mid-August, was $274.

An Inaugural Research Project

One hot and muggy day in mid-August 1937, I trudged over to the Cruft Laboratory. I was told that Professor Frederick V. (Ted) Hunt wanted to see me in his office. I had been depressed all week because I had so little money to carry me through the next year. Should I get a job and take a year off, or try to borrow money—a practice not encouraged in those Depression days? And how long could I keep going on two meals a day?

Nervously, I wondered what Professor Hunt wanted. His specialty was acoustics, a field in which I had no training. An ebullient character, whose brisk stride made it always seem as though he were trying to catch a bus, he was only 32 years old and my height, though already a bit on the paunchy side. His high-ceilinged office in the northwest corner of the building was brightly lit by a half dozen enormous, uncurtained windows.

Professor Hunt swung his chair around, fixed a penetrating gaze on me through his rimless eyeglasses, then broke into a smile. "I'm glad to see you," he said. "I've heard about your last year's course and laboratory work, which have been exemplary. I've just received a grant from the dean of the university that enables me for the first time to hire a research assistant, and Jack Pierce [no relation to G.W.] and Professor Mimno have recommended you. I'd like to explore your interest. I'll tell you more later, but I'll just say now that my aim is to develop a new means for playing phonograph records. I need help right away in setting up a listening studio, including the design of a loud-speaker and the construction of some electronic equipment. I can offer you half-time employment at $800 for nine months. You'll be able to take two courses, instead of the regular four, and I will see that you have a research room of your own."

I was almost speechless. The amount he offered was nearly twice what I had spent the previous year, and my mind summoned up images of three sensible meals a day. And a research room of my own, to boot—almost too good to be true. I thought a few seconds. "Professor Hunt," I replied, "I can say right away that I'll accept your offer. I need the money to stay in school, and I'll try hard to be an effective research assistant." He said he would be back from vacation in a month, and we could begin work then. I would be on the university's payroll as of September 1.

The modern era of acoustics is generally thought to date from the early-twentieth-century work of Wallace Clement Sabine, a professor of physics at Harvard and a pioneer in the field of architectural acoustics. Sabine became known worldwide for the acoustical design of Boston Symphony Hall, a marvel of quality in sound ever since it opened in 1900. In Harvard's physics department, he was succeeded by acoustics specialist Frederick Saunders, from whom Hunt had received his own training. When I signed up for Hunt's acoustics course in the fall of 1937, only vaguely aware at that point of Harvard's rich history in acoustical science, the idea of concentrating in that area—much less making a career of it—had not crossed my mind.

When next we met, Hunt explained that Harvard had made recordings of the Tercentenary Celebration—the event I had happened upon my very first day on campus—on vinyl disks that could not be played back using current phonograph pickups weighing 60 grams, because at that weight the soft vinyl surface would be ruined. His (our) goal was to develop a pickup weighing only a few grams. He had obtained a room for me in the connecting Lyman Physics Laboratory, where I was to set up a playback studio; next to it was a room for my own use. My first job, he announced, was to develop a bass loudspeaker that would give the full range of tones of the lower instruments in a symphony orchestra (he already had a loudspeaker for tones in the higher registers). To do this, I would also have to develop a new kind of power amplifier; he promised to help me with the acoustical principles.

The bass loudspeaker was needed so that Hunt could hear and demonstrate the sound quality from his pickup. Under his general guidance, I designed a folded exponential horn with a mouth opening 5 feet high, 6 feet wide, and 4 deep. For the high-frequency range, the high-frequency multicellular horn was placed on top. Because there were no commercially available power amplifiers to drive this huge loudspeaker unit, I designed an amplifier using the latest large-capacity vacuum tubes. With this equipment, successive designs of Hunt's pickup, and my loudspeaker, the sounds of *Also Sprach Zarathustra* roared down the corridors of Lyman Laboratory every day.

The phonograph pickup consisted of a tiny coil of wire to which was attached a tiny cone with a diamond needle at its apex, all suspended in a magnetic field. The biggest problem was a "ringing" sound, which Hunt and I discussed at length. Every few days, Hunt would modify the pickup, add "damping" materials to kill the ring, and give me a sketch of what he wanted to try next, which I would carry to the machine shop. A day or two later, I would collect the finished device and bring it back to him; after giving it a good listen, we would discuss the next permutation. For a break from our work, we would often head out to the corner drug store. He was addicted to an afternoon fix of soda fountain Coca-Cola, which had a good deal more sugar and caffeine than the familiar bottled version. I stuck to ice cream.

By the end of the academic year, we had achieved Hunt's goal—a phonograph pickup with impressive sound quality weighing only 2 grams. Appearing on the September 1938 cover of *Electronics Magazine,* the new pickup revolutionized the phonograph industry. Whereas the old shellac 10-inch records had only 3 minutes of playing time per side, Hunt's invention made possible the LP—a 12-inch vinyl disk with 23 minutes of playing time per side.

Balancing the Personal and the Academic

Early in June 1938, my sweetheart Floss came to Boston at my urging. I wanted her to see where I was working for my doctorate and to give her some idea of life in New England. Also, I was uncertain I could make it out to Iowa that summer.

I wondered aloud to John and Mabel Barnes where Floss might find lodging. After some thought, they said that she could stay at their home in Medford. With that offer in mind, I planned to call for her each day, take her sightseeing, and drop her off after dinner. To my surprise, John and Mabel said that they'd be away on a trip and we could have the house to ourselves. When Floss came, she and I spent ten days together, reveling in all the tourist attractions of the Boston area. We swam in the ocean, rode the commuter boat, immersed ourselves in museums and movies, strolled around Harvard, the Glass Flowers exhibit, and the Cruft and Lyman Laboratories, and generally had a fun time. But then things took a surprising turn for the worse, and we broke off our relationship soon afterward.

The Big Storm

Then, on September 26, what came to be known as the "Great Hurricane of 1938," category three, blew through New England with peak steady winds of 121 mph. I had been in the laboratory until about 5:30 that evening, when I departed for dinner at Harvard Square. The morning papers had given no hint of an approaching storm. When I stepped out, the wind was whipping all sorts of objects through the air and nearly knocked me over. I sprinted across Harvard Yard to the nearest restaurant. When I emerged an hour or so later, the storm had passed. The roof of the Paine Music Building lay on the sidewalk where I had just passed. The Yard was a chaos of debris; Kirkland Street, which leads to Irving Street where I lived, was blocked by fallen trees and telephone lines. Electric power was off and telephone lines were down statewide, although the university had service because its lines were underground.

Senior research associated Jack Pierce and Professor Harry Mimno mounted an expedition to reestablish some communication between Boston and cities to the west. The Cruft Laboratory had a large panel truck outfitted with a huge bank of storage batteries and an antenna pole at the edge of the roof that could be folded down. Inside the rear door were a radio transmitter and receiver, licensed by the FCC for experimental broadcasting. Knowing of my skill with Morse code, Mimno and Pierce asked me to join them and arranged with the State House to send messages back and forth in Morse.

Seated in front, the three of us headed toward Worcester and Springfield on Route 20 the morning of September 27. Trees and roofs were down everywhere.

We found that we could detour our way around any obstacles, although we sometimes had to go through open fields. We stopped in Marlboro, Worcester, Sturbridge, and Springfield. Local police and relief agencies sent messages through us to the media and the State House. We heard of no storm-related deaths. Most people seemed to consider the event a novelty, except for those whose property had been severely damaged. We spent three nights in motels and returned to Cambridge four days later.

At Harvard, the Second Year

My term as research assistant took a new direction in the 1938–39 academic year. Referring to the recent book by Philip Morse of MIT, *Vibration and Sound,* he had used in his acoustics course, Hunt told me: "Morse has a chapter on sound in rooms, but his work is theoretical and largely untested. I want to explore the limitations and practical uses of this new theory and you will work with me toward that end." I understood the material in Morse's 1936 book nearly as well as Hunt did, having, in effect, explored it together in his class the previous year.

We decided that our first experiments should investigate the behavior of acoustical materials in a model of a rectangular room. My notes read:

We made a chamber of six Transite slabs two inches thick. Two of the dimensions of the box were fixed at 24 and 16 inches and the third could be varied from 13 to 30 inches. The whole chamber was held together by extended steel clamps. We explored the sound field in these chambers, first bare and then with one wall covered with sound absorbing material.

This research resulted in a significant advance on Morse's theories. Hunt published a paper titled "Investigation of Room Acoustics" in the *Journal of the Acoustical Society of America,* formally acknowledging my contribution at the end. The Dean's Fund, he wrote, "made it possible to secure the services of Mr. Leo L. Beranek who has ably executed the experimental details of this investigation." Although some of my colleagues thought he should have named me coauthor, the theoretical material in the paper was Hunt's alone, so I had no objection when he did not.

I visited the home of John and Mabel Barnes several times that fall. Just after Thanksgiving, they had me over for dinner, along with Professor Weston Bousfield, the eminent cognitive psychologist then at Tufts, and his wife, Thelma. At one point, Mabel said: "I don't want to embarrass Leo, but he has broken off with his Iowa girl and is now interested in meeting women." Thelma told me that she had a younger sister, Phyllis, whom I might like to meet. She gave me Phyllis's telephone number and promised to tell her I might call. Which I did, and we made a date for Thursday evening, December 22, to attend a concert by the Boston Symphony Orchestra at Symphony Hall, with the famed violinist Jascha Heifetz as guest artist.

Professor Saunders, meanwhile, had begun to delve into the reasons why violins made by the old masters were so superior to even the best ones made in modern times. He arranged for Heifetz to visit Cruft Laboratory the very day of my date with Phyllis. Heifetz was to play notes on his two famous violins, a Stradivarius and a Guarnerius, and, for comparison, on a violin that Saunders had bought at a pawn shop for five bucks. Saunders asked me to assist him that day. Heifetz bowed individual notes on the three violins for nearly two hours, as we recorded the sounds. I was standing next to his wife when he played a short piece on the "standard of badness" violin. When I told her, "That fiddle never sounded so good," she replied: "He can make any violin sound good."

After the tests, I asked Heifetz about the cheap violin. "I played on a worse violin at a public concert in New Delhi, India," he told me. "The weather was hot and very humid and I had brought only the Strad with me. When I opened the case, I found that the humidity and heat had softened the glue and the instrument was unplayable. Almost in panic, I asked my hosts to find another violin. They came up with an aluminum instrument over which a thin layer of wood had been cemented to make it look like a regular fiddle. I did my best to compensate for its deficiencies while playing it, and after the concert I quickly locked the violin in its case. People came up and said that my Strad had beautiful tone."

That evening, I met Phyllis in the lobby of Symphony Hall. Proudly, at intermission, I recounted how I had spent the morning recording Heifetz at the Cruft Laboratory. Phyllis's eyes widened. I made another date with her, the start of a relationship that ended in marriage.

A milestone worth mentioning, if only by the way: 1938 was the first year in which I filed a federal income tax return. My income for the year was $1,194.00, on which I owed $2.57 in taxes.

"My scientific knowledge," I wrote at the end of the first semester, in January 1939, "shows a great deficiency in mathematics, a general ability to set up a problem and to find out all there is to know about it both experimentally and mathematically, a lack of knowledge in the general field of modern physics, and, last but not least, too little experience in the reading of scientific German." Despite this gloomy assessment, I continued to study German, translated several abstracted articles for publication, read two books on modern physics, explored the use of analogous electrical circuits for determining the performance of electro-mechano-acoustical systems, and completed a study of copper oxide rectifiers. It was a busy time, filled with hard work and the self-criticism that doubtless proved a motivating force in its own right.

The year was made all the better by my association with a Chinese student, Maa Dah You, who was also working toward his doctorate. Maa was about my height, thin with black hair, and spoke with barely noticeable accent. He took everything one said seriously. The Japanese were increasing their incursions into China—the two countries had been at war since the mid-1930s—and every day the newspapers printed stories of rapes and other atrocities. When I asked Maa about his mother and sister, who were still back in China, he told me they had left Peking (Beijing) and were now in Yunnan (Kuming), to the southwest, "where they are protected by the Communist leader Mao Tse-tung." At least once a week, we took the subway to a restaurant in Boston's Chinatown, where I learned to enjoy Cantonese food and to use chopsticks. To this day, Maa refers to me as his older brother—I am four months his elder.

We earned our doctorates together and, in 1941, after Maa had returned home and settled in at the Institute of Acoustics in Peking, he asked me to come over on a one-year appointment as adjunct professor. Much as I would have loved to see him again, with the decade-and-a-half civil war still raging in China, I thought the better of it. Maa tried to persuade me that I had nothing to worry about: the city was surrounded by a immense wall, rendering it impervious to assault by Mao's forces, but I wasn't so sure.

We tried to keep in touch through the years. Maa became one of China's leading acousticians, an expert on special building materials. A devotee of Confucius, he once sent me a volume entitled *The Four Books,* with the note: "Now you know what Confucius *really* say." In the first few decades after the Communist takeover in 1949, I kept writing, but Maa stopped writing back. Finally, a Finnish scientist who had just spent time in Maa's lab told me that, because Chinese officials had started opening and reviewing mail to test levels of loyalty and orthodoxy among the citizenry, particularly within academic and professional ranks, Maa wanted me to stop writing too. During the Cultural Revolution (1966–76), Maa and his family suffered humiliation and loss of livelihood, but fortunately no physical harm. Vilified by his students, he was placed under house arrest for several years; his wife, a physician, ended up at a camp several hundred miles away. A decade later, in the wake of Deng Xiaoping's historic visit to Washington, D.C., in 1979, Maa and I finally renewed contact and—forty years after his original invitation—I paid my first visit to China.

But we are getting ahead of ourselves. As the new semester started in February 1939, Hunt and I planned to measure room acoustics in a full-size rectangular room:

Available in the Jefferson Physical Laboratory was a concrete room with the dimensions 20 × 14 × 8 feet. We started with all walls bare, and then determined how the sound field changed when one of the walls was fully covered with a sound absorbing material. The mathematics was complicated. Fortunately, Maa joined the investigative team and he helped in a major way to develop the theory and to assist in making computations.

I worked long hours on this project and was dead tired by the end of May. Hunt was planning to publish a paper on our research. I suspected he would list himself as the author, with Maa and me mentioned in the "Acknowledgments" as helpers. But I was wrong. "Ted just told me that he plans on putting Maa's and my name on the paper with his," my diary recorded. "God only knows I appreciate it." The paper was published in July in the *Journal of the Acoustical Society of America;* around the same time, Hunt nominated me to membership in the honorary national science society, Sigma Xi, to which I was elected. I paid five dollars for the key.

At Last, a Project of My Own

By now, I had completed most of my required coursework and passed the oral qualifying examination for study toward a doctor's degree, covering the field of fundamental physics. Most important, in my non-Hunt time, I had laid the basis for my doctoral thesis. Recognizing that my work for Hunt required a better understanding of how acoustical materials—such as those used in office buildings—absorbed sound, and that Morse's book offered little practical help, I set about determining "acoustic impedance," an esoteric quantity for measuring the acoustical properties of a material that has no parallel in everyday life. By early June 1939, I thought up a way to measure this parameter:

The measurement of acoustic impedance requires a means for determining the ratio of the sound pressure to the particle velocity perpendicular to the surface of the material, preserving the relative phases. I accomplished this with the following equipment: "A six foot length of Shelby tubing . . . with a rigid termination at one end . . . in the center of which end will be the emitting source . . . and at the other end a moveable piston on which the sample is mounted. Two measurements are made with and without the sample in place: the length of the tube measured at resonance to the surface of the sample and the width of the resonance curve."

When I ended my two years as Ted Hunt's research assistant, I was firmly dedicated to the field of acoustics and immersed in my thesis research.

There were times for relaxation here and there, amid the challenge and excitement of my work. In mid-August 1939, Jack Pierce and his wife, Catherine, invited me to her family's summer home in Brookfield, New York, for a few days. As a senior research assistant in the Cruft Laboratory, Jack specialized in the behavior of the Heaviside Layer, the stratum of ionized air above the earth that makes long-distance radio reception possible. We played croquet and badminton, picnicked, and went to see the movie *The Hound of the Baskervilles,* with memorable performances by Basil Rathbone and Nigel Bruce as Holmes and Watson. In the countryside near the Finger Lakes, I was especially struck by the peat bogs, which covered the marshes in floating mats so thick you could walk on them, and when you pushed a pole into them, it seemed to go down forever. After a completely restful time, we drove back to Cambridge on August 25.

Soon afterward Dr. and Mrs. Roger Hickman invited me to their home in Chatham, Massachusetts, on the "elbow" of Cape Cod. Roger was a technical manager for the Cruft Laboratory and the house was owned by his wife's grandmother. I stayed there for two weeks in mid-September. I rowed, sailed, and drove with Roger to see many parts of the cape. We lingered at the site of the radio transmitting station that Marconi used in 1903 to send code signals from South Wellfleet to England. His radio towers had fallen and were still lying on the sand in 1939; since then, thanks to coastal erosion, the Atlantic has swallowed up the entire site.

Because my visit coincided with the opening salvos of World War II, the Hickman household was all abuzz with news from overseas. Each evening after dinner, we gathered around the radio in the parlor for the latest, wondering what, if anything, could be done to stop Hitler.

After my visits to Brookfield and Chatham, I felt thoroughly rested and ready to move ahead with my research. I also found time, in the period from June to October, to read well over a dozen books by a range of writers, from Lord Rayleigh and Madame Curie to Agatha Christie and Sir Arthur Conan Doyle.

New Opportunities

One day in April 1939, when I was at my tiredest, I got a call from Professor George Birkhoff, dean of the Faculty of Arts and Sciences and chairman of the Committee on General Scholarships. "I am Dean Birkhoff," his voice crackled on the other end of the line. "Are you Leo Beranek?" He told me I'd been awarded a Parker Traveling Fellowship, which would not only cover my full tuition but would also provide an additional $1,125 in spending money. I would be free to travel anywhere I liked while working toward my doctorate, and the fellowship would pay those expenses as well. In a long letter that arrived a few days later, Dean Birkhoff explained:

The holder of a Parker Fellowship is entitled to receive a letter of appointment, bearing the seal of the University, naming the field of study in which he is to occupy himself, and recommending him, as a distinguished student of the University, to the confidence and friendly consideration of the all persons to whom he may present the letter. The [Selection] Committee regards the Parker Fellowships as among the highest distinctions conferred by the University on students; and it trusts every Parker Fellow to make his work worthy of the distinction.

My letter of appointment arrived on April 10, signed by Harvard's president, James B. Conant.

Because travel to research laboratories in Europe's universities was next to impossible during the war, I decided to stay put—to concentrate on completing the requirements for the Doctor of Science in Electrical Engineering and Communication Physics and not to be distracted by other projects. When I explained this to Hunt, he agreed to be my advisor and to serve on my thesis committee.

I didn't have much of a social life that year, apart from my weekly dates with Phyllis, to whom I had taken a great liking, and who was imbued with Boston's rich heritage and culture. We enjoyed going to plays and concerts more than anything else. "Phyllis and I are very happy and have good times together," I wrote on July 1, 1940, "She seems to be a sweetheart and an intellectual friend all in one."

Using the new equipment outlined above, my measurements of acoustic impedance went ahead without a hitch. Nevertheless, at the end of the first semester, Hunt expressed some doubts as to whether my thesis research would be adequate for a doctorate in June; he recommended that I take another year. I listened, but vowed to move faster and, accordingly, I spent every hour late into the night mulling over mathematics, recording data, and analyzing results. By mid-March, I was already busy writing the text of my thesis; by April 1, I had completed most of the drawings. With these in hand, I reported to Hunt that I planned to go ahead with my thesis. When he neither endorsed nor objected to my plan, I completed my thesis manuscript and gave it to the secretary for typing. She finished her work by April 20. I sent out six copies of the thesis volume for binding, and submitted the finished product—"Precision Measurement of Acoustic Impedance"—to the doctoral review committee before April 30. Thinking back on it, I suspect Hunt's tentativeness reflected concerns about his own position as much as they did those about the adequacy of my work. He wanted smooth sailing all the way; any foul-ups by me—his very first doctoral candidate—could be interpreted as a failure on his part.

On April 30, I went to Washington, D.C., to present a paper on my thesis work at the spring meeting of the Acoustical Society of America. Among those present was Professor Philip Morse of MIT, whose book Hunt had used in his acoustics course. Morse seemed to take special interest in my experimental results and asked for a copy of my thesis, which I was only too happy to provide on my return to Cambridge.

My trip to Washington was indeed memorable. The trees were in full leaf, flowers were in bloom everywhere, and the city looked bright and clean, especially when compared to the dismal, down-at-the-heels look of Cambridge in late spring. Taxis were cheap and everywhere. Even office workers took them—two passengers for 20 cents, including tip. My hotel roommate Bob Watson and I went to the top of the Washington Monument by elevator. Then we walked down—all 897 stairs. The calves of my legs were sore for days. I especially enjoyed visiting the Lincoln and Jefferson Memorials and the National Gallery of Art.

The Real Test

Hunt had assembled a gilt-edged thesis committee to give my work a thorough going-over: J. H. Van Vleck (a mathematical physicist—and corecipient of the Nobel Prize in Physics in 1977), E. L. Chaffee, and G. W. Pierce, and of course Hunt himself. To my great relief, the committee approved my thesis some ten days after receiving it. "On Wednesday, May 15, 1940," the announcement read, "Mr. Leo Leroy Beranek will present himself for an oral examination for Doctor of Science degree in Communication Engineering in the special field of Acoustics."

Although a defense of thesis session was normally attended only by committee members and the candidate, under Harvard rules, it was open to the public. My pal Bob Watson slyly decided to drum up an audience, telling Hunt that morning there would be a large crowd. The site was changed to a lecture hall; more than a hundred people showed up. Following my 20-minute presentation with lantern slides, at the invitation of Committee Chairman Chaffee, six members of the audience rose to ask questions, all of which I was able to answer. Then it was Chaffee's turn. He asked me to derive the flow of acoustic energy past a point in space, assuming a free, progressive, plane wave. I had never done this, but I went to the blackboard and wrote out the relevant equations. When my derivation showed "energy passing by a point in the form of 'sausages,' not continuously," the committee didn't know what to make of it. They had never seen this result before; it contradicted their instincts—and mine, for that matter. They argued among themselves. Not finding anything wrong with my analysis, they adjourned the session and retired to Professor Pierce's office to assess my presentation. I waited expectantly in my laboratory room. Finally, later that afternoon, Hunt looked in to tell me I had passed.

During the first half of May, while the committee members were reading my thesis, I was hard at work preparing two scientific papers for submission to the *Journal of the Acoustical Society* (*JASA*). One described the apparatus and the proof that I had made valid measurements of acoustic impedance; the other presented measured data on a number of commercially available acoustical materials. Acknowledging receipt of the papers on June 1, 1940, *JASA* published them both in its July issue, setting the stage for my professional life.

Professor Philip Morse had not seen any data on the acoustic impedance of commercial materials until he heard my paper in Washington. After reading my thesis, he decided to combine my measured data with his 1936 theory. Using some of the same acoustical materials I had measured, Morse and two students at MIT's Eastman Laboratory—Richard H. Bolt and Richard L. Brown—promptly took their own measurements in a reverberation chamber.

Their paper "Acoustic Impedance and Sound Absorption" was published in the October 1940 issue of *JASA*. "Following the excellent measurements by Beranek of the acoustic impedance of various acoustic materials," the opening sentence read, "it is now made possible to demonstrate the adequacy of the theoretical connection between the acoustic impedance of the material and its physical properties on one hand and the absorption coefficient as employed in usual practice, on the other." This was a clear endorsement of my research; it was also an unusual recognition of a freshly minted Ph.D. by the preeminent figure in his field.

Transitions

With my doctorate in hand, my thoughts turned to jobs. Through Professor Pierce, a position came up with the Bell Telephone Laboratories in New Jersey, and through Professor Hunt, one with the General Radio Company in Cambridge. But I chose to stay at Harvard on a temporary appointment as "Instructor in Physics and Communication Engineering on the Gordon McKay Endowment, to serve for one year starting September 1." With a salary of $2,500, I was set for another year of three meals a day.

That summer, I worked for Professor Pierce. "G. W.," was he was familiarly called, had come to Harvard from Texas as a graduate student. A brilliant student, he was appointed to the faculty right after earning his doctorate. He invented the crystal oscillator that enabled radio stations to remain fixed on their frequency—before that, they would drift on the dial and sometimes their signals would overlap. Pierce's oscillator also permitted heterodyning, whereby many simultaneous conversations could be carried over one telephone line, a tremendous saving in wires. For that, the Bell Telephone System paid him handsome royalties.

Pierce and his wife had a cottage in the hills near Franklin, New Hampshire. One day in June, after hiring me, he asked me if I had a car. When I told him public transportation from Harvard Square was so good I had no need for one, he replied: "Let me emphasize that in New Hampshire a car is an absolute necessity. Please go to the Ford dealer in Harvard Square and bargain for a one-seated coupe, I will pay for it." I returned in a few hours to report that I had bargained hard and that the dealer had come down to $750, and even thrown in a set of chains. Pierce wrote a check, which I delivered. Then, euphoric, I drove the car back to the laboratory.

Pierce paid me $300 for three months. He was writing a book on electric circuit theory. "Your duties," he told me, "will be to type clean pages from my handwritten material and to make drawings. You will also do odd jobs for my wife and me, including shopping." Mrs. Pierce was old-style New England, very frugal and well educated. When G. W. had proposed to her, or so the story goes, she said she wouldn't marry him if he asked her to fix their meals. So he hired a cook.

I was given a cabin to live in, on the side of a steep hill about a hundred yards from the main house. Inside were a cupboard, two tables, a desk with bookcase above, a studio couch, a Franklin stove, and, in a small bathroom, a clothes closet and bureau. Outside, by the door, was a pile of wood.

The job turned out to be largely manual labor. G. W. did little writing (he never finished the book), so my typing and drawing duties were minimal. I mowed grass, landscaped, did plumbing, dug a well, and ran errands with the car. I generally got out of bed at 7:45 AM, made my breakfast of coffee, toast, jam, and oranges, and listened to the news. At 9 AM, I appeared at the house to sweep the veranda, trim the petunias, and burn garbage. I worked until noon doing whatever odd jobs G. W. assigned that day. I ate most of my meals at Main Street Lunch in Franklin. Some afternoons I worked, others I drove around looking at scenic attractions—lakes, creeks, and rivers—in the vicinity.

The Pierces' nearest neighbors were the Cannons. Walter B. Cannon was an eminent physiologist on the faculty of the Harvard Medical School. His wife, Cornelia, was a writer with a best-selling novel, *Red Rust,* to her name. They had a son and three daughters. Marion was the youngest. An evening or two each week, the Cannons would invite the Pierces and me over primarily to chat about politics and the war in Europe. Their constant guest was Arthur M. Schlesinger Jr., who was engaged to Marion. The wedding took place that very summer, outdoors on the Cannons' gravel driveway. I remember tying Arthur's bow tie before the ceremony (he didn't know how).

I had a tough time making up my mind about the war. Opinions were divided on whether we should get involved. The most influential arguments came from the America First Committee, a political pressure group—isolationists, essentially—headed by Charles Lindbergh and other prominent figures who opposed aid given to Britain by the United States. I listened to their impassioned speeches and was completely taken in. I remember sitting with guests at the Pierce home when one of them asked me what I thought. I said I favored the position of the Lindbergh committee, and the gentleman shook his head in disbelief.

In my diary, I describe myself as "crude and blundering," perhaps referring to my role as general factotum around the house, but more likely to my occasional forays—ill informed as they often were—into political dialogue. I enjoyed the car and felt really lost after I had to return it to G. W. in Cambridge at the end of the summer.

Among the Pierce's houseguests was a young lady named Frances from a local family, whom I dated several times that summer. We went to movies mainly, on my evenings off. Her family invited me to dinner once and presented me with a book by Agatha Christie. My time with Frances helped make the summer enjoyable, but that was all there was to the relationship. Throughout this period, I was in love with Phyllis Knight, to whom I wrote constantly.

On my return to Cambridge, at the beginning of the 1940–41 academic year, I moved into Conant Hall, right next to Harvard's permanent Glass Flowers exhibition. My roommate, Bob Wallace, was studying for his doctorate in mathematical physics; indeed, he almost seemed to think in mathematics. He was my height, with black hair and eyes that were so weak he had to peer through a magnifying glass when reading. His family owned an important regional telephone company in Texas, and, to show how wealthy they were, he wore a diamond ring. We had a corner apartment, with several windows, a bedroom, and a large living room accessed through double doors, which effectively blocked the sound. We bought an oriental-looking rug and upholstered furniture. When the United States joined the war late in 1941, my draft number was quite high, and it seemed unlikely I would be called up.

The world of electroacoustics was about to open up at Harvard, and I was ready to become a part of it.

3 Wartime: Communications and Kamikazes

When the fall semester opened in 1940, I felt less like a wide-eyed novice. I had come to know the Cruft group quite well, idiosyncrasies and all; my first stabs at research and publication had been well received; a fellowship had come my way; I had slogged through my thesis preparation, earned my doctorate, and been asked to stay on for at least another year. Although the minefield of academic politics was a mystery to me (and would remain so for a long time to come), I knew enough about the various factions to stay out of their way. I wanted to avoid intrigue at all costs and was raring to forge ahead.

My first test came early. Just after I passed my oral exam, Professor Howard Aiken asked me to be a research assistant in a computer laboratory he was just forming. Having heard he was difficult to work for, I declined, then worried whether my decision would make him antagonistic toward me. It didn't seem to, but, like other graduate students, then and now, I was well advised to tread gently around fragile egos. As to the other Cruft professors, I hardly knew Ronald King and had taken just one course under Leon Chaffee in my first year as a graduate student. I felt at ease with Ted Hunt because I had served as his assistant for two years; I had also worked for G. W. Pierce at his summer home in New Hampshire that very summer.

What would I be assigned to do as an instructor, I asked myself? Because my skills lay in acoustics and Hunt would certainly teach that course, it seemed unlikely I would be given a teaching assignment. I would probably be connected to the laboratories in some capacity, even though the other two instructors already running them did not seem overworked.

When I went in to see Professor Chaffee, he told me that three of the perennial laboratory experiments needed to be either replaced or upgraded. On Hunt's recommendation, I had been chosen for the job. Chaffee said the three professors closest to the experiments, which involved new vacuum tubes and electronic measuring equipment incorporating the latest knowledge in radio communications, would discuss with me the lines along which the changes should proceed. My success depended on finding a way to work smoothly with these three; I would also need to coax technical help out of the other instructors.

In what turned out to be a wise move, I set the program in motion only after consulting with the other instructors. I sounded out their opinions first, then shared mine, drawing on my experiences as a student. To my surprise, our ideas meshed without much difficulty or disagreement. I had suspected right off that James Shepherd, an instructor at Cruft for two years already, might covet my assignment as both a challenge and a welcome shift in his own routine, but even he signed off on the agreed-upon program without objection.

I was expected to complete the upgrading in a semester, which seemed eminently doable. Then one week into the term, Professor Chaffee was suddenly hospitalized with a detached retina. Standard treatment in those days meant lying in bed for a month or two with bricks on either side to immobilize the head in the hope of helping the retina reattach on its own. Staring up at the ceiling from his hospital bed at the Massachusetts Eye and Ear Infirmary, Professor Chaffee asked me to substitute-teach his course, with the same material he had taught when I took it.

Now I had a graduate course in electronics to oversee and three laboratory experiments to upgrade at the same time. I would have to commit 12 hours, 7 days a week. Those hellish few months also proved to be a satisfying challenge, however. The students in the course were brilliant; having done little research on the topic, I had to prepare myself thoroughly to ensure I was on top of the material for each lecture—no off-the-cuff delivery would do. Chaffee's notes saved me, and there were no complaints (at least none reached my ears).

Three times a week, I arrived at Cruft by 8:00 AM to teach Chaffee's course. After walking to Harvard Square for a noontime bite, I was back in an hour, ready for a long afternoon of work. Supper in the Square was followed by more work. Much to my relief, Chaffee returned in early November. By that time, I had completed the revision of two of the three experimental setups, earning praise from Professors Mimno, King, and Chaffee for my speedy results. I was just starting in on the third when the flow of events swiftly and irresistibly swept me along.

MIT physics professor Philip Morse continued to be intrigued by my work, having read my thesis and my two papers in the July 1940 issue of *JASA*. To me almost a god, Morse had written a textbook on sound that was radically different in its call for a new way of thinking about and measuring the behavior of sound in rooms and of acoustical materials, such as those installed on the ceilings of offices and on walls to optimize sound effects. I was taken by his expressive face, which sported a mustache and well-trimmed beard, and by his readiness to respond to questions, and felt refreshed every time I talked with him.

My contacts with Morse had always been cordial and businesslike. Then, out of the blue, in mid-September 1940, he wrote me that my data "seem to clinch the whole room acoustics discussion. There is a great deal more to be done. . . . One possibility is that you and I write a joint paper giving more details of the theory and giving detailed experimental checks." I felt excited and overwhelmed at the same time. I told him how interested I was, but that I had to complete the laboratory changes assigned to me at Harvard before undertaking anything new.

A New Federal Project

Sometime in October, Professor Morse called with an entirely different proposal, to say an acoustics project was starting up that he thought might interest me. The project was sponsored by the newly created National Defense Research Committee (NRDC), whose mission was to conduct defense-related scientific research using qualified civilians. Karl Compton, president of MIT and a prime mover in the early days of the NDRC, had told him that the Army Air Corps (the Army Air Force was formed a year later) urgently needed a lightweight, efficient acoustical material to reduce intense propeller noise in their bombers, which it believed was causing pilots to become overfatigued. Compton had asked Morse to set up a project to take on this task. Morse thought of me because of the close relation to my thesis work. Then came the clincher: he wanted me to come to MIT as his right-hand man by the first of the year.

I jumped at the opportunity. I felt confident that I could finish my laboratory commitments and, with Harvard breaking for the usual January "reading period," that the move would not inconvenience my students. I assumed Harvard would be happy to release me for this defense effort.

Professor Hunt was anything but happy. In fact, he was furious and did not want me to leave. The National Defense Research Committee had already given MIT money to set up the Rad Lab, as the Radiation Laboratory was commonly called. There was no earthly reason, he fumed, why MIT should be awarded every federal project that came along. "I will go to Karl Compton and tell him that the project should be at Harvard, that *I* should direct it, and that you should be *my* right-hand man." Sure enough, he went over to MIT to rail at Compton, who apparently then discussed Hunt's objections with Morse. But to no avail—Morse wouldn't budge. Hunt then threatened to go to Washington and formally object to MIT's taking on the project. Detecting intransigence on both sides, Compton called the two rivals into his office and said, in effect: "A plague on both of your houses—the work will be done at Harvard and Beranek will be the leader." To keep himself in the mix, Morse suggested that the NDRC set up a joint project with me in charge of the experimental work at Harvard and with him in charge of the theoretical work at MIT. This became the project's modus operandi, although no theoretical work ever came out of MIT.

Compton then proposed a supervisory committee, the Committee on Sound Control, with Morse and Hunt as chairman and vice chairman; with Harvey Fletcher, director of acoustics research at the Bell Telephone Laboratories, and Hallowell Davis, professor at the Harvard Medical School as members; and with me as secretary. Morse then held a brief planning meeting in the faculty room of the Physics Department at Harvard, where he, Hunt, and I talked over our next move. There was no hint of the recent conflict—no apparent ill feelings either—and I was asked to prepare a year's budget to develop the new acoustic material. We also scheduled the next meeting, which was to be attended by the full committee.

In the meantime, I had concluded I would need an assistant and some additional equipment to develop the new acoustical material. As director of the Cruft Laboratory, Professor Chaffee said that Harvard would continue to pay my salary and that he could provide space for my assistant. No mention was made of overhead. I estimated that the funding for the assistant and additional equipment would cost about $1,500 and $2,500, respectively.

A Budgetary Surprise

The committee met formally for the first time in New York on November 8, with members of the National Defense Research Committee and the Army Air Corps also in attendance. The twelve of us sat in a circle, with no table between us, in a large conference room in the offices of the American Physical Society. Morse introduced us one by one.

In due course, when Morse moved on to the budget, I proposed that $4,000 be appropriated for the year. There was an awkward silence. Somewhat irate, Captain Fred Dent of the Air Corps stood up and moved that the problem was so urgent as to require a twentyfold increase—on the order of $40,000 per six-month period. No one argued with this, and I was authorized to spend $80,000 for the year.

That sum was a shocker, equivalent to about 2 million current dollars. Annual salaries for technical personnel then were $1,500 to $3,000, and there was neither withholding tax nor health, death, social security, and retirement benefits to be paid; also, overhead fees at universities were much lower than they are now. How would I put all this money to good use? The NDRC representatives said that in three weeks a contract would be available for Harvard to sign and that we should get to work right away. The committee recommended that I visit the large aircraft companies as soon as possible to learn what was already known about reducing airplane noise. It also voted to set up a parallel laboratory to investigate the effects of noise on human behavior—fatigue, psychomotor efficiency, and the like—with a director to be determined later, and with the same hefty budget. Within a month, this effort was established at Harvard, with Professor S. Smith (Smitty) Stevens as director. Smitty was given space in the basement of Memorial Hall, where he established what became known as the "Psycho-Acoustic Laboratory" (PAL).

I remained somewhat dazed by the rapidity with which events were unfolding, the urgency of our mission, and the vast sum involved. At the time, there was only one telephone for the entire Cruft Laboratory. This prized instrument was held in the central secretary's office and served eight professors, various instructors, and a number of graduate students. To summon a person to the telephone, a wired network had been set up with a miniature loudspeaker in each room. Everyone was assigned an alerting signal—mine was "dit-dah-dit-dit," Morse code for "L." If the Air Corps was to get its answers as promptly as it had demanded, this simply wouldn't do. Riding with Hunt on the train back to Boston, I blurted out: "Ted, I must have my own telephone in the laboratory right away." To which he replied, "Now, Leo, take it easy."

Harvard's Reaction

Not surprisingly, I did not accept Hunt's advice. Instead, within days, I arranged with Professor Chaffee not only for a separate telephone but for a separate office as well. Two days after that, I went directly to the financial vice president of Harvard, John W. Lowes, whose office was across the corridor from the president's in Massachusetts Hall. Impatient to find out why I was there, Lowes waved me to a seat at a small table. After explaining the purpose of our government project, that it had to start right away, that I would be the sole director under an outside advisory supervisory committee, and that a contract would be available for signing within a month, I asked him how I should proceed.

A modest-sized man, whose desk, clothes, waxed mustache, and carefully trimmed eyebrows reflected an impeccable neatness, Lowes explained that this was the first research funding that had ever come Harvard's way from the federal government. He rocked back and forth in his chair for a minute, twisting the ends of his mustache, then quickly announced that, because of the military nature of the project, he would set up an account at the Bursar's Office called "Anonymous Research under Leo Beranek." I was to keep a simple ledger with income on one side and expenditures on the other. To cover salaries and other expenses, I was to send vouchers to the Bursar's Office, which would in turn dispense payments. With that, he wished me the best of success and suggested I come back if any problems arose. When Harvard's treasurer, William Claflin, discovered this unusual, secretive agreement three years later, he was appalled at the lack of fiscal control and set up a formal bookkeeping system with appropriate checks and balances.

Next, I had to lay out the project's needs for Professor Chaffee. I estimated that over a dozen scientists would be needed, along with secretaries and assistants, offices, space for research equipment, and access to Cruft's machine shops. He agreed to provide space at Cruft and additional space as necessary in the adjoining Lyman Laboratory of Physics. Cruft would bill me for services in the machine shops. Space was available, Chaffee said, because a number of Harvard's physics and engineering professors had moved to the Rad Lab at MIT. He offered to set up security gates and whatever else might be required, with overhead payments equal to 50 percent of all salaries. The NDRC accepted these terms, no questions asked.

Assembling a Staff

My next step was to hire the right personnel. Under the recently passed Selective Training and Service Act, every able-bodied man between the ages of 21 and 36 was subject to the draft and to military service in sequence according to a lottery number. Those engaged in key government-sponsored research projects and certain educational activities, however, were granted draft defer-rals. That being the case, I was able to get graduate-level acoustics students to sign up with a guarantee of just one year's salary. Ted Hunt gave me the go-ahead to approach his students; I also asked the deans of engineering at the Universities of Michigan, Minnesota, and Texas for the names of students with training in acoustics.

I then went to see the head of MIT's Rad Lab, Lee DuBridge, a fellow alumnus of Cornell College in Iowa. After chatting about our alma mater, we turned to the subject of war research. DuBridge told me the Rad Lab planned to hold a recruiting session at Harvard's Jefferson Laboratory—just across the way from Cruft—where the lab's director of personnel, Wheeler Loomis, was to make a presentation. Somewhat presumptuously, I asked for and got permission to speak as well. Loomis outlined the Rad Lab's tremendous needs, which made it seem almost as though my relatively small operation was hardly worth considering. Undaunted, I began with "That reminds me of a story." And I proceeded to tell about a meeting of America's railroad owners, where the owner of a railroad that was only a few miles long demanded the same voting rights as all the others, arguing that his tracks were just as wide as New York Central's. The anecdote helped pique students' interest in my project.

My first hires, who went on to become the senior members of my staff, were Francis Wiener and Robert Wallace from Harvard; Rudolph (Ruddy) Nichols from the University of Michigan; Harold Ericson from the University of Min-nesota; and Wayne Rudmose, Robert Newman, and Sparky Ennis from the University of Texas. They were all about the same age; only one, Rudmose, was married and therefore not so interested in evening work. I also hired an older man by the name of Henry Jaffe to handle the details of settling staff in at Cruft, assigning offices, and helping them find housing. I still had one laboratory experiment to finish revising and, with help from Bob Wallace, completed the series by Christmas, as originally planned.

Off to the West Coast

During the three weeks before Christmas, I visited six principal manufac-turers of military aircraft: Douglas, Boeing, Consolidated, Lockheed, North American, and Curtiss-Wright. As we landed in El Paso, one of our refueling stops (transcontinental flights of the day were unable to fly nonstop coast to coast, I was handed a Western Union Telegram, which read: "Call Jones NDRC Washington. Ur trip funds not yet authorized." Jones (the perfect name for a quasi-anonymous government official) informed me that my project would be approved in another day or so and that I should stay over in El Paso until I got the go-ahead. As anticipated, the permission took just one day, but, while waiting, I crossed the U.S.-Mexican border to Ciudad Juárez. Getting back turned out to be not so simple. The U.S. Immigration Service demanded my draft deferment card. I had never even heard of the need for such a document, a reflection of the pressure and rapidity of events as well as my total immersion in the project at hand. The immigration officials finally let me back in, but not before issuing me a stern warning to get my deferment status in order. I telephoned Professor Morse, who had a deferment card waiting for me when I got back to Cambridge.

In the search for new noise-measuring equipment, I contacted a friend at the ERPI Company in Los Angeles, leading manufacturers of audio instrumentation. He said that, if I could spare a few hours, he would show me around the MGM studios, which were quite interesting from an acoustical standpoint. In one of them, we had the good luck to run across Lana Turner, Hedy Lamarr, Edward Everett Horton, and L. B. Mayer himself all hard at work on *Ziegfeld Girl*—no sign, though, of Jimmy Stewart or Judy Garland.

Later, as I headed over to the Douglas Aircraft Company in downtown Los Angeles, I saw that a huge camouflage net, on which fake trees and shrubs were mounted, had been placed over the factory buildings. The idea was to confuse Japanese pilots should their commanding officers decide to send them on a cross-Pacific raid. Inside Douglas's main building, I was ushered into the engineering department and introduced to two young engineers who had been in charge of quieting the DC-3s used by most commercial airlines at the time. They gave me a great deal of information, including several internal reports. The company's engineers were excited by the prospect of learning about any new materials and methods.

I attended brief meetings, too, at Boeing in Seattle and at Consolidated Aircraft in San Diego. Boeing's focus was more on the military end of things; they manufactured the B-17 bombers used on bombing missions flying out of England, and their engineers confirmed that cockpit noise was so intense that voice communication proved nearly impossible, especially at high altitudes. Consolidated was supplying the Navy with pontoon aircraft designed to land on water. All six companies agreed to let us measure noise levels in their aircraft, and tentative arrangements were made for visits in January to the Consolidated, Douglas, and Boeing plants.

The Work Starts

Back in Cambridge, I met with my new staff and talked about how we should proceed. Wayne Rudmose, Bob Wallace, and Ruddy Nichols, who had the most experience, helped plan our next steps. I proposed that we begin by understanding noise conditions in the pilots' compartment. To do this, we needed to take actual noise measurements in existing aircraft for comparison with noise measurements in ground vehicles. Using information from Douglas and Boeing, we then needed to determine which acoustical materials would provide the most noise reduction and how and where these materials might best be installed in the airplanes. Most important, we needed to develop a new, more efficient acoustical material. Douglas had suggested that the best acoustical material was Kapok, a fiber from the pods of a plant, but it was clear that Kapok was ill suited for use in military aircraft because of its combustibility.

Our immediate task, however, was to procure the best possible instruments for measuring noise. Once we compared what EPRI in Los Angeles and General Radio in Cambridge had to offer, it became clear we should buy our instruments from EPRI. I asked Wayne to make procurement, and then he and Sparky Ennis should head out to Consolidated, Douglas, and Boeing in January to take aircraft noise measurements. Bob and Ruddy were to be responsible for devising equipment to measure acoustical materials and for determining how the materials should be mounted in aircraft for the best results. Calling everyone's attention to the new Psycho-Acoustic Laboratory (PAL) being built by Smitty Stevens in the basement of Harvard's Memorial Hall, I stressed that we should give him whatever assistance we could with planning the test rooms and with selecting loudspeakers for use in testing the effects of noise on human behavior. In the next few days, I would work with Henry Jaffe to set up security gates, hire a secretary, and acquire office equipment.

We were off to a good start, but many challenges remained. Although I had experience working with acoustical materials, I had never given any thought to developing a material different from what was already commonly used in buildings. At every opportunity, I tried to get my mind around what might constitute a lightweight, highly effective, but incombustible acoustical material. Here are some of my thoughts at the time:

For maximum sound absorption, one should have as much surface area inside the material as possible so that the air particles have lots of area on which to rub. Now, consider an acoustical material made of circular fibers. Ask, "What happens if the fibers are made smaller and smaller in diameter?" The smaller the fibers, the more fibers can be crammed into the acoustical material. But, the remarkable result is that as the diameter of a circular fiber is made smaller, its weight decreases much more rapidly than the surface area. This means that if the acoustical material is made from very small fibers it can be made very light in weight and at the same time for that weight it will have the maximum surface area inside, thus absorbing more sound than the usual commercially available materials.

I then considered how existing acoustical materials were produced. Most came from wood shavings, or cornstalks, or mineral flakes. The Kapok material used by Douglas Aircraft was made from small natural fibers that were lightweight because hollow. Heat-insulating materials were made from circular glass fibers and were available commercially, but the fibers were solid and large, hence too heavy for aircraft. I called the Owens Corning Fiberglas Company. Could they make an acoustical material from tiny diameter glass fibers? They had no idea how, but said they would try.

Just two months later, Owens-Corning called to say it had produced glass fibers one-tenth the diameter of a human hair by squeezing glass through extremely small holes—and could send us a one-pound sample for testing. Our tests on this new material confirmed my calculations. Owens Corning went on to produce a new lightweight acoustical blanket, called "Fiberglas AA," which is widely used in passenger aircraft to this day.

The first report of our project appeared in July 1941. We wrote that we had measured noise in airplanes made by several manufacturers and that by comparing the size of engines, propellers, and fuselage construction, we were able to determine why one airplane was noisier than another. We also described the new acoustical material we had discovered. But our report also pointed out a major obstacle—that there was not much room for acoustical material in the pilots' compartment because of the large windshields and bulky electronic equipment mounted on the sides. The Navy Bureau of Aeronautics did not seem discouraged, however, and wrote us how pleased they were with our rapid progress. The Army Air Force responded in like manner and urged us to work with the airplane companies to achieve whatever noise reduction we could.

My life was not all work, although it often seemed that way. I took Phyllis out most Saturday evenings and several times went with her to Trinity Episcopal Church in Boston on Sunday mornings. She loved going to concerts, plays, and movies, and we enjoyed many of these together. From time to time, her family, the Knights, who lived in Jamaica Plain, just outside Boston at that time, invited me over for a Saturday dinner of Boston baked beans and brown bread, which I found quite tasty. One time, Phyllis, her family, and I traveled by overnight boat to New York City. I spent two days there, and returned by overnight boat. Phyllis knew the city and took me sightseeing. In the evening, we all went to Radio City Music Hall to see the stage show.

"Eureka!" Moments

Because it was impossible to find spots to place significant amounts of sound-absorbing materials in aircraft cockpits, and because in-flight communication was nearly impossible owing to high noise levels, Bob Wallace came up with the idea for a "noise-canceling headset," which would improve speech audibility for pilots and crew alike. I presented this proposal at the meeting of our supervisory committee on April 25, 1941. In a follow-up letter to the Navy Bureau of Aeronautics, I observed that Bob's "pressure-canceling device" (as we called it) would enable signals to reach the ear while providing large reductions of external noise. We proposed using Smitty Stevens's facility to test whether the observed communication failure resulted from outside noise interference at the ears of the listener. If this was the case, it would be highly desirable to develop a device like Bob's.

Smitty, Ruddy Nichols, and I went to Washington to sell the idea. The Navy, however, was more interested in upgrading its present equipment than in developing complicated, new earphones that might not be put into service for a long time, if ever. In that vein, a formal directive arrived in September from the Chief of the Bureau of Aeronautics of the Navy, asking us to start a confidential research project aimed at improving speech intelligibility of communication components under simulated flight combat conditions. We were to concentrate on upgrading amplifiers, headsets, and various types of open and mask-enclosed microphones, even oxygen masks.

The Navy directive set the course of our Research on Sound Control Project, as it was formally called, and that of Smitty's outfit—the Psycho-Acoustic Laboratory—for the next two years. My immediate mission was to establish a research program with the help of my senior staff, hire extra personnel, and secure additional office and laboratory space in the Cruft and Lyman

Laboratories. More electroacoustic equipment had to be purchased and Bob Newman's high-altitude test facility had to be replaced with a new one four times as large, so that we could measure the properties of microphones and earphones at low atmospheric pressures, similar to those encountered in unpressurized, high-altitude bombers. Our staff quickly doubled in size.

To gauge the intelligibility of speech heard using military audio equipment in loud noise fields, PAL set up experiments in which the speaker would keep the loudness of his voice constant while listeners wrote down the words they could hear through the equipment. PAL staff discovered that the only young male subjects they could find for daily intelligibility tests were conscientious objectors who had declined to go to war or to help the war effort. These young men agreed to work on this project because the results were expected to be useful in peacetime.

As activity in our parallel research projects ramped up in July 1941, I proposed to Phyllis—and she accepted. Surrounded by a small group of family and friends, we were married on September 6 by a Presbyterian minister in the chapel of Emmanuel Church, Boston. Bob Wallace was my best man. My Iowa family couldn't come, being tied up with preparations for my brother's wedding a few days later, back home.

While I was dressing for the ceremony, a special delivery letter arrived. Far from being a timely message of congratulations from some well-wisher, however, it was an angry letter from Frances, the young woman I had dated in New Hampshire in the summer of 1940. She demanded that I cancel the wedding and "come back" to her. In no way had I misled her that summer, had not communicated with her since, and she had no reason to think otherwise because we had never been intimate.

Happily, the wedding went off without incident. I pushed from my mind all images of Frances showing up just as minister got to "speak now or forever hold your peace." Afterward, Phyllis and I stood at the chapel entrance and chatted with guests as they filed out. We then joined them for a modest reception at the Knight home. After a first night at the Parker House in Boston, an upscale hotel in those days, we spent the rest of our honeymoon at the Hotel Frontenac in Quebec City. Our stay there was a delight, thanks to the attentive management. We savored the superb cuisine and enjoyed a special suite at a very reasonable price. I still have a keepsake from the trip—a menu with the footnote "Room service 5 cents extra." Phyllis and I explored just about all of Quebec City and loved it so much that we returned several times in later years.

Back in Cambridge, we rented a one-bedroom apartment in a building that was just opening up at 50 Follen Street. Our fourth-floor flat had an entry hallway, large living room, bedroom, kitchen, and bath. All for $60 a month.

Lunching with Eleanor Roosevelt—Twice

In Washington, D.C., after talking over some matters relating to the Harvard contract, I headed out to the airport the next day—November 27, 1941—to catch a late-morning flight to New York. The last to board, I was shown to the only seat available, near the rear of the DC-3. This aircraft was equipped for sleeper service, which meant that sets of two seats faced one another on each side of the narrow center aisle. Opposite me sat a woman hidden behind a newspaper. After I was belted in, she lowered her paper. Taken completely by surprise, I stammered out, "Hello, Mrs. Roosevelt." Mrs. Roosevelt returned my greeting and resumed reading. A short time later, she folded her newspaper and said: "I hope you will excuse me—I have many letters here that I must handle." I nodded. She took an already opened letter out of her bag, perused it, wrote a short note on it, put it back, and went on to the next.

About twenty minutes out of Washington, the stewardess came by and placed a table between us to serve lunch. Mrs. Roosevelt put aside her letters and we chatted as we ate. She wanted to hear my views on the war in Europe. "Many people," she said, "accuse Franklin of being a warmonger and ask him to turn instead toward peace. He tells them that they should go talk to Adolf Hitler, who is the real war monger." By then I could sympathize, having come a long way politically in a short time from my naive acceptance of the line pushed by Lindbergh and other isolationists. Mrs. Roosevelt soon changed the subject to Christmas. She asked if I had any idea what she should get for her two sons serving overseas. I suggested books, sweaters, neckties, pen and pencil sets, but none seemed to take her fancy. She then asked me what I was doing and wanted to know about my family. As our plane made its approach, she said she was late for a noon luncheon downtown. Which was better, she wondered, to go into the terminal and find a telephone or rush right out and get a taxi? I offered to call ahead while she headed for the cabstand. She thanked me and wrote a telephone number on a slip of paper.

When the airplane pulled up to the gate, someone said a car was waiting to take Mrs. Roosevelt into the city. She turned to me and said, "Please write me a letter saying that we were on this flight together, and I will invite you and your wife to the White House." I walked away a bit stunned, then sped toward the phone booths.

Back home in Winchester, I asked Phyllis if she could guess who I had just had lunch with. "I suppose," she responded playfully, "with Franklin Roosevelt?" "No," I beamed, "but with Eleanor." I carefully drafted a letter to Mrs. Roosevelt, dated and mailed December 6. The next day, a Sunday, Phyllis and I went for brunch at the home of one of the Naval officers enrolled in graduate work at Harvard. Four officers were there with their wives, along with most of the Cruft faculty and their wives. About noon, someone telephoned to tell our hosts that Pearl Harbor had just been attacked. We gathered around the radio, dismayed. One of the officers said, "This means we will be called to active duty tomorrow." The party quickly broke up.

I expected to hear nothing more from Mrs. Roosevelt. As preparations for war got under way in earnest, she was appointed by her husband as codirector with Mayor Fiorello LaGuardia of the nation's civil defense. A letter dated December 12 arrived from Mrs. J. Helms, Mrs. Roosevelt's secretary, apologizing: "I am sure that you will understand that due to the seriousness of the national situation and the increased demands upon her time because of official civilian defense work, Mrs. Roosevelt has had to make many changes in her calendar." Then, in a second letter from Mrs. Helms dated December 20, came the surprise: "Mrs. Roosevelt hopes that you and Mrs. Beranek can lunch with her at the White House on Friday, December twenty-sixth, at one o'clock." I wired our acceptance.

Phyllis and I took the sleeper train to Washington on the night of December 25 and registered the next morning at a hotel within a few blocks of the White House. Shortly after noon, we hailed a taxi and asked the driver to take us to the White House. He said there was no admission there and we could not enter the grounds. We asked him to try anyway. When the cab stopped at the guard booth on Pennsylvania Avenue, we showed our letter of invitation. The guard checked the list of approved guests and waved us through. The door was opened by the butler, who welcomed and escorted us to the Red Room. A White House attendant came by to say that Mrs. Roosevelt was at the Capitol attending a joint session of Congress addressed by Winston Churchill, and she should be back about 1:15. One other luncheon guest waited with us. Meanwhile, sitting in the visitor's gallery of the Senate chamber, the Roosevelt party listened with rapt attention, along with much of the nation, as the eloquent British leader, in a historic speech, forecast hard times ahead—

but ultimate success—for the Allied Forces. At the end, he flashed the "V" for victory sign to thunderous applause.

Mrs. Roosevelt came to the Red Room somewhat flush with excitement and took us to the family dining room on the first floor, adjacent to the formal dining room. Positioning herself at the center of the long side of the elliptical dining table, she seated me to her right and Phyllis across from her, to the right of Harry Hooker, a friend of the Roosevelts from Hyde Park. She explained that the president was dining in the Oval Office and expected Soviet Ambassador Andrei Gromyko to join him. She also mentioned that Prime Minister Churchill was a guest in the White House and had asked to retire to his room after his energetic performance at the Capitol. Someone advised me before we sat down that Mrs. Roosevelt was hard of hearing and that I should speak up.

The meal began with soup, followed by a meat dish. Mrs. Roosevelt asked her guests what they thought of Churchill's speech. Those who had heard it thought well of what Churchill had said. Mrs. Roosevelt remarked that, in her opinion, the German people did not want war, but Hitler had cleverly maneuvered them into it. She hoped Americans would be sympathetic to Germans when the war ended. Not everyone, of course, shared her confidence about the eventual outcome.

I was fascinated by her style of dining. A small cup of mixed nuts stood in front of each place setting. She took a slice of bread, laid it on the table by her plate, spread butter on it, placed nuts in precise rows on the slice, cut it in half, and ate it. We had water, but no wine. After each guest had weighed in on Churchill's speech, Mrs. Roosevelt told us of her trip to the West Coast, as codirector of Civil Defense, to help ease concern there about the possibility of a Japanese invasion. In a remarkably open and straightforward way, she held LaGuardia to task for not joining her on the mission. "I tell Franklin," she confided, "that LaGuardia has too many jobs and should not have been appointed to this position." Then she called special attention to one of her guests, Mayris Cheney, a professional dancer whom Mrs. Roosevelt had appointed to take charge of a national fitness program. "I believe," she said, "that in this critical period the American people should be asked to exercise every day to keep healthy. I am suggesting that they exercise from six to seven in the evening, which will give them an hour to dress for dinner at eight." I felt like interjecting that the Americans I knew did not dress for dinner, nor did they eat at eight, but I bit my tongue.

Just before dessert, the waiters brought each guest a plate with a finger bowl on top, a spoon on one side, and a fork on the other side. I did as the others did—set the finger bowl off toward the middle of the table and placed the silverware on either side of the plate. Then a tureen of cooked blueberries arrived. Mrs. Roosevelt suddenly reached over and picked up from the middle of my plate a thin crocheted doily about three inches in diameter, lifted my finger bowl, and placed the doily underneath it. I must have thought the doily was part of the plate design. "You may think this belongs there," she scolded, "but it doesn't." The party broke up soon after we consumed the blueberries. Mrs. Roosevelt accompanied Phyllis and me to her elevator. We were joined by a White House attendant who, Mrs. Roosevelt said, would give us a tour of the main floor. We thanked her warmly and said our farewells.

Full Steam Ahead

Back at the Cruft Laboratory, communication equipment—microphones, amplifiers, and earphones—kept flowing in from the Army Air Force, Navy, and Army Ground Force. The earphones or headsets were usually mounted in aviation helmets. All equipment went first to Nichols and his staff, who thoroughly measured physical properties. When those tests were complete, it went to the Psycho-Acoustic Laboratory for articulation testing—where words were read over the system to a group of listeners who wrote down what they heard.

The conclusions reached by the two laboratories were startling. The proportion of words that PAL subjects heard correctly in simulated aircraft noise using standard service equipment was less than 60 percent. The tests carried out in our laboratory revealed that the primary villains in this loss of intelligibility were the earphones themselves. We detected a marked lack of strength at the higher frequencies, which meant that words with voiceless consonant sounds such as "th," "sh," "t," and the like were unrecognizable. We rushed our findings to the military. On reviewing them, our Navy liaison officer grew even more concerned and thought Smitty Stevens and I ought to go to Washington to present our data. He asked us to appear Friday, February 13, 1942, at a meeting of a newly formed Joint Radio Board (JRB) established under a joint Army, Navy, and British military communications directive. The JRB had been given full responsibility for communication equipment in the Army Air Force, Air U.S. Navy, and British Royal Air Force.

The February 13 meeting was the Joint Radio Board's first, intended to set an agenda. Instead, we *became* the agenda. I can remember the chairman asking Smitty and me to present our findings. I stood up faster, so went first. Smitty followed with the results of PAL's articulation tests. We told the board that our tests showed, in no uncertain terms, the need to procure new earphones as soon as possible. They took copies of our draft report for study and asked us to return four days later for further discussion.

At the second meeting, on February 17, we were asked to seek out every design of earphone in the country that might be a candidate for replacement, with the primary criteria that it must fit into existing Air Force, Navy, and British aviation helmets and also attach to the headbands of the U.S. Army and Navy. Improvements in the microphones, helmets, and oxygen masks could follow. Together, one of the board's engineers, Frank Tennenbaum, and I were to visit manufacturers of earphones, who were asked by the military brass to give us full cooperation. We were given the highest priority on airline flights, essential in those days when commercial passenger flights were far less frequent than they are now.

Frank and I visited seven companies. Only three had earphones small enough to fit into military helmets, and none of their instruments incorporated magnetic shields to protect navigation equipment. Inadequate though they were, I took the three best models back to Cambridge for study. Our performance testing showed that only one—a unit made by Western Electric Company—would be adequate for our purposes. We came up with specifications based on this model, emphasizing, however, that other companies would be at liberty to compete for contracts if they could produce earphones of similar size and performance.

Most critical now, a protective magnetic shield had to be designed. Just one month after the JRB had approved our program on Sunday morning, March 22, a group of officers from the four services met in my office at Cruft to decide on the design of the metal case. The helmets of the three air services were laid out on a table and I had a craftsman present to make sketches. By the end of the day, we had the design of a case that would hold the Western Electric headset, that would fit into all helmets, could be clipped into Army and Navy headbands, and would be encased in steel to contain its magnetic field.

On April 10, the Army Air Force and Navy Bureau of Aeronautics issued a directive that all headsets under current contract were to be abandoned and our new headset design adopted. Western Electric and the Permoflux

Corporation were the only two companies to submit final samples. We crash-tested these in our and Smitty's laboratories. Both designs fully met our audio intelligibility requirements. Western Electric's headset was designated "ANB-H-1" and Permoflux's "ANB-H-1A."

But our tests were not yet complete. The Air Force wanted us to perform articulation tests at high altitudes in unpressurized aircraft. This would require about ten enlisted men as subjects, with everyone wearing oxygen masks. Almost right away, we were directed to the Army Air Force Base at Wright Field, Dayton, Ohio, to plan the tests. We reported to Colonel Benjamin Chidlaw, who scheduled a flight the next morning so that we could study the problems involved in altitude tests under real operating conditions.

Weather delays were quite common at Wright Field, we quickly learned, and could stretch to a week or more. Thus it wasn't until the third day of trying that we finally got a flight. We had at least three days of flight tests in mind so that, counting on-ground preparation time, our work might take a month or more, which we found unacceptable. We headed over to Chidlaw's office to ask if there was anywhere else we could go. To our surprise, front-desk staff told us he had just been promoted to Brigadier General and had already left for the Pentagon. Somewhat frustrated, we returned to Cambridge.

Sunday night, I took the sleeper train to Washington. At the Pentagon, with some difficulty, I found Chidlaw's new office amid a maze of rooms and the nonstop hustle-bustle of wartime. He greeted me like a long-lost friend, adding that, with his telephone not yet hooked up, he had unlimited time to discuss our problems. After I recounted the weather difficulties at Wright Field, he said he would arrange for us to go to Eglin Field in Pensacola, Florida, where the weather was almost always good, assuming we were lucky enough to avoid the occasional hurricane. Smitty and I flew down the next day to size up Eglin.

The commanding officer sat behind a desk in a large, hot, unair-conditioned room. We sat facing him. Chidlaw had telephoned him with orders to look after us properly, but we were hardly shy about laying out our needs. We would require, we said, housing for our staff—up to six people—and a room for our equipment. He said our group would be welcome to stay in the officers' quarters and that we could have use of a room on one side of a hangar for technical purposes. On Chidlaw's orders, he provided a group of enlisted men for articulation tests. In the meantime, another weather-related problem cropped up. The humidity in Pensacola was so high that the standard condenser microphones used for calibration would quickly become inoperative. When I returned to Cambridge, I arranged for an air-conditioning unit the size of a small refrigerator to be shipped down for installation in our assigned equipment room.

Four of us arrived at Eglin with the new interphone equipment, a microphone calibration apparatus, and repair tools. A day of preparation was followed by two days of successful articulation tests at high altitudes, with decompression sickness (the "bends") affecting only one of us. But on the afternoon of the second day of in-flight tests, the commanding officer called me into his office. Until written orders for the work arrived from Washington, he told me, we could not continue. I thought a second and said: "Your office is unbearably hot—if you let us complete our tests tomorrow, we will have our air-conditioning unit brought to your office for you to keep." Not surprisingly, we got the go-ahead for our final day of work.

By the late fall of 1942, just one year after the United States entered the war, the ANB-H-1 headset had become standard in the aviation services. Our two Harvard laboratories had gained a reputation for accuracy and speed. We had made friends at the highest levels in the Army Air Force and U.S. Navy and gained the esteem of the highly regarded research staffs of Western Electric and Bell Telephone Laboratories. This reputation would stand us in good stead later.

Another Priority Request

Somewhat earlier, in the summer, another priority request had arrived from U.S. Army ground forces. They asked for help in evaluating military loudspeaker systems. The Army believed that powerful sound systems could be mounted on trucks, which, when deployed during landings on beaches, could be used to give soldiers oral instructions—or, alternately, on land for deception to simulate tank and truck noises at battlefield locations, for example. Because of the intense sound levels such systems produced, they could not be tested outdoors in inhabited areas. A large chamber had to be built and lined with an acoustical material that would replicate the outdoor environment, which is to say, that would reflect no sound back to the loudspeakers—and very important, that would allow no audible sound to escape. I designed an apparatus to test acoustical structures for lining the chamber's surfaces. A senior at Harvard College then ran tests for me on nearly a thousand different structures. The one that best absorbed sound over the widest frequency range was a glass fiber wedge 8 inches square and 46 inches long, separated 11 inches from the concrete wall behind it.

The chamber that we built had interior dimensions of 38 feet by 50 feet by 38 feet high, with walls 12 inches thick. It was erected on Oxford Street, just across the street from Harvard's Glass Flowers. Because wedges had to be installed on all six surfaces, access to the interior of the chamber came by way of a 4-foot-wide "railway" track suspended from steel cables 12 feet above the floor. Nineteen thousand wedges were needed, requiring nine American railway boxcars to bring the glass fiber material from Ohio to Cambridge, where they were fabricated in a temporary "factory," which I set up in a vacant warehouse with the aid of Oliver Eckel, a local businessman of remarkable ingenuity.

We completed the chamber in record time. Our measurements showed the wedge-covered surfaces reflecting hardly any sound and nothing could be heard outside. The first loudspeaker tests were carried out in the summer of 1943. The chamber became known in the Harvard community as "Beranek's Box." I coined the term "anechoic," meaning "without echo," and, after I used the adjective in a published paper, it eventually (1962) appeared in *Webster's Third International Dictionary*. Since then, thousands of anechoic chambers have been built around the world, most using my pioneering wedge structure or a slight modification of it.

Several months earlier, toward the end of 1942, I happened to be watching a newsreel in a movie theater and observed that our soldiers needed to remove their metal helmets in order to use their communication systems. How about coming up with very thin earphones, I wondered, thin enough to fit underneath the helmets? I studied one of the new ANB-H-1 earphones with this prospect in mind, and realized that by rearranging the magnets, its thickness could be reduced from one inch to three-eighths of an inch. I had a sample put together in the shops, and our measurements showed that it worked as well as the original. I carried the unit down to Washington and showed it to the procurement people in the War Department. They were surprised and immediately contracted with the Western Electric Company to manufacture tens of thousands at the highest priority level.

By late fall 1943, our work had migrated almost completely away from noise control in aircraft to speech communication systems. We changed our name to Electro-Acoustic Laboratory (EAL) in parallel to Psycho-Acoustic Laboratory (PAL). My title was changed to director of EAL.

On April 10, 1944, the secretary of the Acoustical Society of America notified me that I had been chosen to receive the society's Biennial Award for Note-worthy Contributions to Acoustics. This award, now known as the "R. Bruce Lindsay Award," is given to someone under 35 years old. A high honor for a young professional, it was presented to me at a banquet in New York City on May 12. I was thrilled when my old mentor and colleague Ted Hunt showed up to read the encomium. I received a fancy certificate and a check for $100.

Incidentally, the fund known in the Harvard Bursar's Office as "LLB—Anonymous Research" totaled $1,213,000 over the course of the war. Taking into account that the salaries of top engineers are now about 30 times what they were then, and that overheads are now twice as high, this sum is equivalent to about $40 million today. And this was by no means unusual—just one illustration of the vast resources that our country was channeling into military programs at the time to meet the global crisis.

From Airplane Communications to Navy Ships

Our work at the Electro-Acoustic Laboratory attracted widespread attention, including that of the Coordinator of Research and Development for the Navy. In the spring of 1943, the Navy brass asked the National Defense Research Committee to fund the EAL and PAL to improve communication systems in the combat information centers (CICs) of Navy ships, particularly with respect to equipment, sound engineering, and selection and training of operators. On each ship, the CIC is a room into which flow streams of data from radars, radios, and other sources. The data are analyzed to detect naval and airborne targets, to determine their course, speed, and identity, and to pass this information expeditiously to ship's command, which in turn mobilizes antiaircraft or surface batteries for action.

We plunged headlong into this work. Within the 1943 calendar year, not only had we thoroughly studied voice communication in Navy ships, we had also issued reports on the design of six new ancillary pieces of equipment for the combat information center, including plotting boards and switching equipment. The Navy was so pleased that the Commander in Chief's Office asked us to undertake a comprehensive study of CIC operations over and beyond communication and acoustics.

Jumping way ahead, when the war in Europe ended on May 8, 1945, I began to wonder about the need for further research on communication systems. I met with senior EAL staff and shared my suspicions that the war in the Pacific might be over sooner than we thought. "Harvard," I said, "has started reseeding the lawns in the Yard after four years of neglect. President Conant knows what's going on and I've heard rumors that a new weapon is in the works to bring the Pacific war to an early end." Some inkling of what a number of our scientists were up to at Los Alamos had managed to leak out in certain quarters despite the heroic measures taken to keep it under wraps. In any event, the purpose of my May meeting with EAL staff was to give them a heads-up on planning their postwar futures. They took my advice and all had new jobs by the fall.

Neutralizing the Kamikazes

Although our Navy had clearly demonstrated its superiority over the Imperial Japanese Navy, a new Japanese weapon threatened to reverse that, sinking the first U.S. ship—an aircraft carrier—on October 21, 1944. By January 1945, kamikazes had taken a number of ships out of combat—sinking some, damaging many, and tying up the Navy's repair facilities. Born of desperation, the kamikaze suicide bomber had the potential to shift the momentum of the war in Japan's favor.

Even before the kamikaze threat, some of us had already foreseen the need to improve operations of the combat information centers (CICs) on our Navy ships. The CIC, in fact, represented the sole means for early detection of an aircraft's attack path. Based on development at the Electro-Acoustic Laboratory, sometime in the late summer of 1943, I suggested that new equipment be evaluated at the Navy's training center on Saint Simons Island, Georgia. The Navy expressed enthusiasm for the proposal and recommended that an advisory committee of five investigators be established—the CIC Survey Group—and that the group undertake familiarization cruises aboard Navy vessels. Harvard was awarded a contract for this work, with investigators to be selected from MIT's Radiation Laboratory, the Systems Engineering division of Bell Telephone Laboratories, and Cruft Laboratory.

The CIC Survey Group consisted of Leo Beranek of Cruft, as chairman, and A. Tradup and L. A. Yost of Bell Telephone Labs, R. W. Blue of the Rad Lab, and Wayne Rudmose of Cruft, as regular members. Our first task involved visiting several Naval training stations on the Eastern seaboard to acquaint ourselves with the broad problem, to learn about the various types of radar equipment used in the Navy, and to evaluate how this equipment operated in the CICs.

Then, in December, Yost and I boarded the heavy cruiser USS *Canberra* for a shakedown voyage. T. E. Caywood and R. L. Wallace took a course in aircraft interception at Saint Simons Island. In January 1944, Rudmose and Blue went to sea on the aircraft carrier USS *Wasp.*

For the shakedown cruise, I had to buy a Navy officer's uniform and acquire collar pins bearing the letters "OSRD" (Office of Scientific Research and Development—a funding entity created by the NRDC in 1940). The uniform requirement was necessary so that, if any of us were captured, we would not be considered "spies." I spent some time reading about traditional organization, command, and mores aboard a Navy vessel. Parenthetically, as far as I know, we were the only ones in World War II to don a Navy uniform for such a short time and return to civilian status right afterward.

I arrived at Newport News on the morning of December 3. Stepping into a restroom at the airport, I changed into my temporary uniform and proceeded to the ship. Chills went down my spine as I walked up the gangplank of the USS *Canberra,* saluted the officer of the deck, and showed my Navy orders. A sailor escorted me to a senior officer, who removed one copy of my orders and said I should present myself to the captain later in the day. He had a sailor show me to my bunk, my home for the next four weeks—an upper berth in a large below-decks space. An hour later, I presented myself to the captain, who, though very busy, greeted me politely. He told me there were 2,500 crew and officers on board.

Having no officer rank, I was permitted to eat in either the officers' dining room or the enlisted sailors' mess, as I wished. I learned over meals that some officers were yearning for early action in the Pacific; others hoped never to see combat.

Anchors were weighed about 6 AM on December 4. As the ship moved down the channel toward the Atlantic, the fog grew so thick we could see nothing beyond a hundred feet. I went to the combat information center and watched the radar screens, which showed the receding shoreline, as well as buoys and other floating objects. The ship gained speed, traveling at around 15 knots, by my estimate, and the crew took no notice of anything in the way except for buoys. We were at war, and any fishermen we encountered would just have to get out of our way. When we cleared the channel and entered the open sea around Cape Hatteras, the vessel began to heave. Many sailors hung over the rails, throwing up. I felt a little queasy, but managed to keep down my breakfast.

I spent much time in the combat information center, the primary reason I had come aboard. The center of attention was the two radar screens, one for air targets and the other for surface targets. Aircraft appeared as blips on a cathode ray tube. When a blip appeared, it was assigned by the top CIC officer to a sailor who called out successive position coordinates from which the aircraft's speed could be determined. The data were marked on a vertical Plexiglas board by another sailor, who stood behind it. The delay between sighting an airplane and determining its position and course from the path on the board was thus very long; voice communication with the gunnery stations added further delay. The whole, cumbersome process was far too slow to bring down many kamikazes—if any at all. We talked with crew members, both officers and enlisted men, who understood a little of what we were up to, but viewed us as some sort of curiosity. We grew friendly with the captain, who was quite naturally anxious to improve the performance of his CIC.

The day before Christmas, the *Canberra* docked in Port of Spain, Trinidad. The officers headed for the Queens Hotel, and we joined them in celebrating Christmas Eve in the main lounge. I sat with junior officers and bought drinks all around. Soon, I discovered that six rum and cokes, bought by others, in front of me, where they sat untouched. When an ensign struck up a medley of Christmas carols on the lounge piano, we all sang along, many of us homesick for friends and family far away. Sometime after midnight, we strolled (some stumbled) back to the dock, where launches returned us to the *Canberra*. The next day, we started our trip back north.

That first night of sailing was under a clear and moonless sky. Having to go on deck through two successive doors made me acutely aware of the risk of emitting even small amounts of light with German submarines lurking nearby. Our captain took us on a zigzag course to make it harder for a German torpedo to strike our hull. Two destroyers, doing their best to follow our erratic course, shadowed us to help shield the vessel from attack. At one point, the *Canberra*'s engines throbbed loudly as propellers were thrown into reverse to avoid colliding with one of the destroyers. I watched a gray hulk slide by just off our bow.

By now, it was clear to me what direction our project should take. We needed a suitable shore location at which to set up a model combat information center, complete with radar, to simulate what actually happened at sea. Airplanes to raid the "ship on land" could come, I hoped, from a Navy air base. Time-and-motion studies were essential, too, since so much depended on effective coordination of the actions of personnel within the CIC.

I returned on December 29; our two investigators on the *Wasp* returned by the end of January. We spent two months analyzing our recordings and another two weeks writing our report, although, delayed by conflicting opinions among the CIC Survey members about where to go from here, the final draft did not reach the Navy until May 24, 1944. After hashing out our differences and reaching a consensus, we recommended what we considered to be the optimal strategy for systems planning and equipment evaluation—simulating a combat information center on land, preferably at a site in or near the Cambridge area because our key research personnel lived there and could not be resettled wholesale to a new location.

Captain Beard, of the Commander in Chief's Office, apparently did not think much of our proposal: we never heard from him. Not long after we submitted our report, however, the Shipbuilding Division of the Bureau of Ships issued orders to set up a project for finding ways to shorten CIC operational times. The Navy was on the verge of awarding a contract for this study to a well-known commercial psychology research center when a rumor surfaced that the company had come under investigation by Senator Harry Truman's Special Committee Investigating National Defense.

The Systems Research Laboratory

In mid-August 1944, about three months after the Navy received our report, Captain C. G. Grimes, head of Interior Communications in the Navy's Bureau of Ships, asked whether the Electro-Acoustic Laboratory would take on the project. Philip Morse of MIT, chairman of our Supervisory Committee, and I went to Washington for discussions about extending our CIC contract. We decided, with Grimes, that a separate laboratory ought to be created. Morse so informed the Office of Scientific Research and Development (OSRD), our source of funding, and the Systems Research Laboratory (SRL) was established under my direction.

I met the following week with Grimes's chief assistant, Lieutenant (later Lieutenant Commander) Robert Bookman to talk over options for where we might set up the Systems Research Laboratory. A number of possibilities on both coasts came up. I kept insisting, however, that the laboratory be located in southern New England. It, I also argued, had to be near the coast and must have a reflection-clear radar screen.

A few days later, Bookman called to say that he and Grimes had found an ideal location: the grounds of the Naval Radar Training Facility on the tip of an island south of Jamestown, Rhode Island, called "Beavertail Point." Set on Rhode Island's south coast, the site allowed for a clean radar sweep of 120 degrees. Airplanes from the nearby Quonset Navy Air Base could be commissioned to "raid" the simulated combat information center. Research personnel wouldn't need to relocate from Massachusetts; if not exactly the easiest of commutes, travel by train or automobile from Boston to Jamestown was manageable enough.

Bookman wrote requisitions for $2 million ($40 million in today's terms) of Navy CIC equipment. One day in January 1945, I got a call from him saying that he, Grimes, and Captain Beard had just solved an almost intractable problem. Standing regulations of the Commander in Chief's Office stated that the latest radar equipment must go directly into the fleet—no requisition would be approved for delivery to a land facility. The ingenious solution was to commission the Systems Research Laboratory as a Navy ship, the USS *Beavertail.* This name was not to be used publicly, only as a shipping destination via the Quonset Navy Air Base in Rhode Island.

As already mentioned, the reason kamikaze aircraft were so successful in striking our fleet was that the time between detection of planes flying at low altitudes and the arrival of information at the weapons stations was so long that guns could not be positioned in time to shoot down aircraft before they reached the ship. As a result, it was said, just about the only effective defense against a kamikaze was a marine on deck with a rifle. Suicide bombers did not need to sink a ship to knock it out of combat; destroying or severely damaging the superstructure that contained the radar equipment and the captain's bridge would most often require the ship to return to a land-based facility for repair.

The most devastating stretch in the kamikaze campaign came with the battle for Okinawa, from April 6 through June 21, 1945. Nearly 30 of our ships, mostly destroyers, were sunk by kamikazes. In addition, 217 ships were damaged; 43 of these were total losses, and 23 required up to several months in port for repair. Suddenly, the System Research Laboratory became vital to the Navy's efforts. Construction of the "ship on land" started March 1, 1945, and was completed June 1 at a cost of $93,000. The structure contained a combat information center space (25 feet by 40 feet), a control room (15 feet by 40 feet), offices, shop, storage facility, and electrical and heater rooms. Two tall towers, 74 and 54 feet high, rose overhead with radar antennas placed on top. I became, in effect if not in name, the captain of the Navy warship USS *Beavertail.*

The staff I pulled together included 7 applied psychologists, 13 physicists, 4 time-and-motion engineers, and 4 operations experts. Two of the time-and-motion engineers came from Purdue University, one from Harvard, and one from New York University. Four consultants were also brought in: two from the Rad Lab and one each from the Bureau of Ships and OSRD. I hired an operations manager, Robert Morton, who tended to everything that I could not handle. I also hired a financial manager, drivers, and messengers.

But where would our thirty or so personnel find temporary housing? The Navy facilities could handle only a few and there were no hotels in Jamestown. Morton learned of a country club building, about two miles away, that had been closed a few years earlier. We took a look and decided to rent it. A crew was hired to clean, paint, and buy new mattresses and bedclothes.

The Systems Research Laboratory began full operations on July 1, 1945. Our combat information center was equipped with the latest radars, communication equipment, plotting boards, and switches. Inputs to the radar screens came either by way of actual airplanes flown out of nearby Quonset Navy Air Base or from three "trainers" developed at the Rad Lab. The trainers were capable of simulating simultaneous attacks by 4 aircraft and 5 surface targets. The time-and-motion engineers could then experiment with ways to speed up the flow of data from radars to gunnery. Owing to the design of the floors in the CIC, that is, 20-inch squares above 12-inch-tall supporting stanchions, equipment could be moved about the room without rewiring, making it a straightforward matter to shift equipment from one location to another.

As the kamikaze onslaught continued, four segments of the Navy—the Bureau of Ships, the Bureau of Ordnance, the Bureau of Aeronautics, and the Commander-in-Chief's Office—viewed our work as their last-ditch chance to reverse the alarming successes racked up by Japan's waves of suicide bombers. The Japanese could build thousands of kamikaze aircraft quickly—the planes needed to be just sturdy enough and to carry just enough fuel to reach their targets and they would have no trouble enlisting suicide pilots as their homeland came under the mounting threat of invasion. Indeed, it was rumored that our invasion of Japan, planned for that November, would have to be put off until Navy ships could neutralize the kamikaze attacks and until the vessels damaged by them were repaired and returned to the Pacific. Our limited defense capability against kamikaze assaults, coupled with the decreasing availability of Navy ships because of those assaults, may have been a factor in President Truman's decision to drop the atomic bomb; at least this was the rumor.

Initially, I spent five days a week at the SRL. About August 1, I rented an apartment in Jamestown and Phyllis joined me. Imagine my surprise when a call came in from Washington to notify me—almost at the very moment it happened on August 6—that an atomic bomb had been dropped on Hiroshima. But, the SRL studies were making such strides and because there was no assurance even then that the war would soon end, Navy personnel went ahead with a planned trip to visit us. They appeared on August 9, which happened to be the day the second atomic bomb was dropped, this time on Nagasaki. A contingent of fourteen officers showed up at the lab in full regalia: a vice admiral, 2 rear admirals, 8 captains, 3 commanders, 2 lieutenant commanders, and 4 lieutenants. Three top executives from Harvard also showed up. Our guests left the laboratory excited and anxious for more results. But, for all practical purposes, the war ended three days later, and everyone's focus turned almost overnight from wartime strategy to peacetime transition.

The War Ends

Phyllis and I were at Jamestown on August 12, the day the Japanese war effectively ended and three weeks before Japan's official surrender. We caught the ferry to Newport to join in the festivities. When we docked about 4 PM, we found the city going wild. Fire engines raced up and down sounding sirens, people rushed about cheering and embracing, fireworks lip up the sky. We ate at Newport, took the ferry back to Jamestown, and the next day I went to the lab with mixed feelings—delighted that the war was over, of course, but disappointed that the Systems Research Laboratory had gotten off the ground too late to make much of a difference. I arranged with Robert Morton to keep the facility going and asked any members of the staff who would consent to remain to continue experiments and prepare reports on their findings. Their reports were completed by the end of the year. The Navy then arranged with Johns Hopkins University to continue our efforts to improve the combat information centers. I was hired as a consultant and returned to Beavertail several times the next year. A year after that, the Navy closed down the base and moved the outfit to Baltimore, thus ending my involvement.

I was exhausted by the time the war ended, so Phyllis and I took a short vacation in New York City. We had a mint julep every afternoon at the hotel—often enough, in fact, that we came to be greeted familiarly as "Mr.

and Mrs. Mint Julep." Then I returned to my position as a Faculty Instructor at Harvard.

Near the end of September 1945, I applied for and was awarded a John Simon Guggenheim Fellowship. Phyllis and I lived through 1946 off the tax-free $2,500 that came with the award. From February through September 1947, I had a contract with the Office of Naval Research to write the book *Acoustic Measurements,* which covered the many new things discovered or learned during the war years at the Electro-Acoustic and Psycho-Acoustic Laboratories. That fall, I moved my office a few miles down the river to take a position as Associate Professor of Communication Engineering at MIT, the next step in a career that had whisked through the academic world to government-sponsored research to service in the Navy—and back again.

4   MIT, Teaching, Writing, AT&T, and Traveling

October 1946. The Red Sox had won the American League championship and were playing the Cardinals in the World Series at Sportsman's Park in Saint Louis. I was glued to the radio in our apartment at 50 Follen Street in Cambridge. It was the ninth inning of the seventh game. When the phone rang, I toyed with the idea of making a rude remark about calling at exactly the wrong moment. But caution prevailed, thank goodness. "This is Karl Compton at MIT," came a quiet voice on the other end. "We would like to talk with you about joining our faculty." Suddenly, I forgot all about pinch runner Paul Campbell poised hopefully to race home and tie the game for Boston (in vain, as it turned out).

President Compton's call came at a critical juncture in my career. I had just been offered a full professorship at the University of Notre Dame, where I was slated to take charge of establishing an acoustics research program. Two days before, I had returned from South Bend, Indiana, where the heads of the Physics and Engineering Departments and several faculty members had welcomed me warmly. I had gone so far as to indicate that I would probably accept the offer. Now I wasn't so sure.

The following Monday I trekked over to MIT to meet with Compton. He offered me an associate professorship of Communication Engineering, and an office in a newly formed Acoustics Laboratory. This was the fulfillment of an aspiring engineer's dream: a professorship in the top technical university in the United States and perhaps the world. But I still debated with Compton, observing that there were disadvantages to teaching in a large institution, that I would not be in command of MIT's acoustics program, and that I had been offered a much larger salary elsewhere. Compton sweetened the offer. I would be technical director of the Acoustics Laboratory on the same level as the administrative director, with a promise of early tenure and a better salary. I went home, thought it over, then called Notre Dame to say that I had decided on MIT.

Getting Started

A few days later, I arranged with Administrative Director Richard (Dick) Bolt to visit the Acoustics Laboratory. I knew Dick through meetings of the Acoustical Society of America, and held him in the highest regard. He had earned his doctorate in acoustics at the University of California in June 1939. A tall, handsome, dynamic man, he was a popular lecturer who inspired many promising MIT students to go into acoustics. He showed me around the laboratory, a converted two-story garage with plenty of research space, a library, two lecture rooms, and many offices. He told me they had construction plans for my office, but first wanted my approval. Which I readily gave, to a space 18 feet square off the entrance corridor and across from his office, with a relatively high ceiling and windows on one side. I asked him about graduate students and was impressed by their number and by the variety of subjects they were working on.

I next went to see Professor Harold Hazen, chairman of the Department of Electrical Engineering. He had discussed my coming with his senior faculty members, and together they had decided I should teach a course on acoustics covering wave propagation, which would be required of all electrical engineering students. And would I also like to work as one of six teaching assistants in Professor Ernest Guillemin's course on electric circuit theory? I jumped at the chance, knowing it would let me delve into some of the Rad Lab's wartime work.

In September 1947, I started teaching acoustics at MIT three times a week to some 160 students crammed into one large lecture room. Much of my material was new. On Monday of the second week, a student interrupted my lecture. "We don't understand what you're talking about," he said in an accusatory tone. "What's the problem?" I asked. "The way you are handling electrical circuit details is different from what we have been taught." Apologizing for not making my approach clearer, I told him I combined electrical, mechanical, and acoustical elements in one circuit (an approach that traditional methods did not permit). At the next lecture, I handed out a table outlining how to combine elements and take duals using my approach. No one complained after that, but the concept was novel enough to keep the students on their toes. Out of this course came my successful textbook *Acoustics,* published in 1954.

My unorthodox approach led, among other things, to the development of those small "shelf-size" loudspeakers found everywhere today. Major Edwin Armstrong had invented a new type of radio signal, called "frequency modulation" (FM), that offered higher quality audio and freedom from atmospheric static. In 1947, FM radio had not yet caught on, one reason being that high-quality loudspeakers at the time were oversized and expensive. My approach, which combined the electrical, mechanical, and acoustical elements of a high-fi system into one circuit, stimulated the thinking of two of my students, who went on to develop small loudspeakers that could reproduce low tones just as well as most larger speakers, although they required larger power amplifiers. After that, FM broadcasting began to flourish. When I stopped teaching acoustics at MIT in 1958, Amar Bose, founder of the Bose Corporation and developer of the popular "wave systems" and related advances in listening technology, took over the course and used my book *Acoustics* as a reference tool.

Outside Ventures

In the fall of 1947, while teaching at MIT, I also became a consultant to the General Radio Company of Cambridge, Massachusetts. I spent every Friday afternoon over there for nearly three years. I helped design three instruments—one for measuring the strength of a sound, a second to separate the sounds into eight bands from low to high tones, and a third to calibrate the other two. I tried to convince General Radio's management that more instruments were needed, but they were against expansion. They felt that the United States would suffer an economic depression after the war—failing to take into account that we were the only country to survive the war with its manufacturing facilities fully intact. General Radio was soon to be outdistanced by the more aggressive Hewlett-Packard Corporation in California.

In 1948, the Office of Naval Research (ONR) was underwriting surveys of research in various branches of physics in Europe. Because of my experience directing the Electro-Acoustic Laboratory at Harvard during the war, they asked me to carry out a survey of acoustics. In mid-July, I sailed from New York to England aboard the Cunard Line's *Britannic.* On my arrival in London, I reported to the American Embassy, Office of the Naval Attaché, at Grosvenor Square, where I was given ration cards, still required in England during those early postwar years for the purchase of basic supplies. I spent 12 days in London and 2 or 3 days each in Delft, Paris, Marseille, Bern, Munich, Göttingen, Copenhagen, Oslo, Göteborg, Stockholm, and Helsinki. I traveled mostly by air. In each airport, I was met by a U.S. Navy officer, whisked through customs, and taken in a car to my hotel. In most cities, the Navy had a dinner the first evening to which local scientists were invited. I visited the top acoustical laboratories in each city and came away with a fairly complete picture of what was being done in the acoustics field in Europe.

In Paris, I joined a group of leading acousticians from around Europe for a day in the research department of the French national telephone company. Another day, I toured the research department of a technical university. I took buses and subways to the usual tourist sights, including a climb to the very top of Notre Dame Cathedral and out onto its sloping roof. One of my MIT students, Bill Lang, had a summer job at the Paris office of the Phillips Corporation and he took me to the top of the Eiffel Tower and to the Louvre. What a thrill in those days to stand six inches from an unprotected *Mona Lisa* and to admire it at leisure.

Phyllis had sent me a photo of the gorgeous view of the rear of Notre Dame from Tour d'Argent, a first-class restaurant she described in glowing terms. So I invited Lang and a local couple I had met to dine there with me. Before leaving the hotel, I cashed traveler's checks for twice as much as I thought the evening could possibly cost. Even so, I had the sinking feeling it was still not enough when the doorman took one look at my tip and threw it into the street. The restaurant was on an upper floor, with a relatively low ceiling, tables just properly separated, and the lighting low so that one could enjoy the scene through the floor-to-ceiling windows. The electricity for the illuminated rear side of Notre Dame was said to have been paid for by the restaurant. I ordered a seven-course French dinner and chose as the entrée their famous roast duck (No. 188179), an excellent wine, a wonderful dessert, and coffee. After the meal, we were taken on a tour of the wine cellars, with some bottles covered by dust a century old. The evening would have cost about $1,200 today, and my available cash just covered the bill. I walked back to the Hotel de Crillon because I had no money for a taxi.

Switzerland was the most beautiful country I visited, especially Bern, where I stayed in the elegant Hotel Bellevue Palace. My windows looked out on the

Aare River 200 feet below, low mountains on the other side, and beyond, the Bernese Oberland peaks. One night, I was a dinner guest in the beautiful home of Willi Furrer, the director of research for the Swiss Telephone Company. The living room walls featured paintings of the Dutch Old Masters. Willi later took me to his laboratories, where I learned of new electronic equipment, which we subsequently purchased and brought over from Switzerland for our laboratory at MIT.

A month later my European trip ended in Helsinki. Paavo Arni, a vice president of the national radio corporation, whom I had met in England, took me in hand and we did the city up and down. At his instigation, I was interviewed by seven Finnish newspapers and appeared the next day on the front page of five. We spent a half day with the distinguished architect Professor J. S. Siren, and visited his famous parliament building.

On the voyage home, once again aboard the *Brittanic,* my cabin mate was none other than Walter Cronkite, who was returning from a two-year stint as a United Press reporter in the Soviet Union. This was more than a decade, of course, before the start of his storied career as CBS News anchor. We had long talks deckside and over meals about the Soviet Union, its secrecy, and the prospects for its economy. I asked him whether graduation from an ivy league university or MIT gave someone a leg up in his field. He thought not, he said; a person's abilities and hard work were far more important. Back in New York, Phyllis and I spent two weeks vacationing before I returned to my fall teaching duties.

The Academic Circuit

MIT's Acoustics Laboratory was booming along, and I enjoyed working there. Starting with 35 faculty, students, research assistants, and staff in 1947, we had nearly doubled in size by 1951, with about 90 percent of our financial support coming from the U.S. Navy. I held seminars in a conference room for a number of graduate students, including several from England, Sweden, Germany, Iran, and elsewhere—usually sponsored by Fulbright funds—encouraging them to ask questions and, when I did not know the answers, finding out by the next day's session. The seminars often generated spirited discussions. I found time to undertake basic research projects as well, sometimes alone and sometimes with one or two graduate students. In one project I discovered that if a ventilation duct was properly lined, one could achieve an unusually high sound reduction in a very short length of duct. I used this unexpected finding later in two consulting projects: the quieting of a Convair airplane and the construction of a NACA muffler.

Then, in the summer of 1951, Professor Hazen asked me to take temporary charge of Professor Guillemin's circuit theory course. The class numbered some 30 students, who had completed a year of basic electrical engineering courses. The second week, I invited them to our house in suburban Winchester for an outdoor buffet party. This gave the students an opportunity to get acquainted with each other outside of class, and they bonded well. Throughout the course, I encouraged them to ask and answer questions and gave weekly quizzes to keep them motivated. When the last class was held, each student handed me a letter of thanks and, together, they presented me with a five-pound box of chocolates. I sent the letters on to Professor Guillemin, thanking him for the year I had spent as his assistant.

At this time I was working on my book, *Acoustics,* seeking ways to make the material accessible. Halfway through the year, one of my students, Mary Anne Sommerfield, offered to help me clarify the material in the first two chapters on the wave equation and its solutions. The book greatly benefited from her suggestions. Other students also helped out. Among those who especially stimulated my thinking, and who read and commented on one or more chapters, were Kenneth Stevens, Jerome Cox, Norman Doelling, and Jordan Baruch, all of whom went on to become leaders in acoustics.

In my first acoustics class, 1947–48, I had two students from Argentina's Navy department: José M. Oñativia and Rogelio R. Alcantara. On their return to Buenos Aires, they convinced Professor Vignaux, director of the Instituto Radiotécnico at the Universidad de Buenos Aires, to invite me to teach acoustics in their winter semester. I replied that, though I was very interested, I could not leave until the MIT semester ended on June 1, and had to be back in Cambridge in time for the start of fall classes in mid-September. Vignaux agreed to these conditions. Phyllis, my toddler son, James (Jamie), born July 8, 1947, and I arrived in Buenos Aires at 1 AM on June 12. José and Rogelio were at the airport to greet us.

The next day, José took us apartment hunting. We found a furnished place above the Hotel Residencial Arenales, at Calle Arenales 3173, for $190 per month plus a monthly service fee of $22, which included all meals in the hotel dining room, maid service, linen, heat, electricity, and water. Our allotted expense account gave us $150 to spend on other things.

The dean of the faculty of the Ciencias Exactas has announced by postcard in early June that I would teach a postgraduate course on "Electroacústica" Tuesdays, Thursdays, and Fridays at 7:20 PM, from June 21 through September 8. The course was to cover fundamentals, physiological acoustics, electro-mechano-acoustical circuits, loudspeakers, microphones, acoustical measurements, and architectural acoustics. Each class would be two and a half hours long. I counted 42 students at the opening session, all of whom came from the graduate school or from industry. When, however, they learned that lectures would be delivered in English—which the dean's announcement failed to mention—some left after 15 minutes, joined later by others who also found they could not overcome the language barrier. As a result, final enrollment fell to 20—still a good-sized group. I spent about 8 hours preparing each lecture, which (written out longhand) I would then pass along to the institute staff, who would translate and put together handouts for the students. It was wintertime in Argentina and so cold in the unheated university buildings we could see our breaths. The students wore coats and I sported a heavy sweater.

Phyllis and I loved opera, but found that opera performances at the famous Teatro Colón, where seating was by subscription only, were completely sold out. A month later, when I told an English-speaking architect how much we would like to go, he arranged a meeting with the director of the Teatro himself, whom we found at the opera house dressed in tails and striped pants. He spoke good English and we talked about acoustics and opera. He promised to call if any tickets became available.

Returning to our apartment in the late afternoon a week later, Phyllis and I were astonished to find a message that ten tickets had been reserved for us at the performance that very evening. With help from the hotel telephone operator, we rushed to call friends and, by chance, four couples were able to go. Most surprising of all, our seats were in the president's box at the center of the first balcony. We saw *Madama Butterfly* that night and later—also with the director's help—*La Traviata*. The performances were of the highest quality, equal to any at the Metropolitan Opera in New York. We were taken on a tour of the scenery facilities, enormous rooms below the adjoining street, exceeding anything we had ever seen in the United States. The opera season, we were told, was financed by an automatic tax, a small percentage of Argentina's income tax. We attended one symphony concert in the Teatro Colón, with tickets purchased on the black market.

My most memorable accomplishment in Buenos Aires was giving a comprehensible lecture in Spanish. The Institute of Radio Engineers, Buenos Aires Section, invited me to speak in a large auditorium at one of their meetings. My talk, "Design of Broadcasting Studios," was some 50 minutes long. About a hundred people showed up. I wrote the speech in English; I had Rogelio translate it and then read it onto a tape recorder. I practiced the speech in Spanish by playing the tape a few phrases at a time, doing my best to mimic his cadences and articulation. By working hard at this for many days, I got the tempo, emphases, and pronunciation to the point where, after the speech, several people said I sounded fluent. When, however—with brash overconfidence—I tried to read a translation at a later institute meeting *without* a tape rehearsal, my listeners could barely understand what I was saying. So I never tried that again.

On our way home from Buenos Aires, we stopped over for a wonderful three days in Rio de Janeiro, staying at the Regente Hotel, a block off Copacabana Beach. We saw as much as we could, even taking a countryside excursion to the Quitandinha Hotel in Petropolis, where President Truman had stayed on his visit to Rio. We marveled at the palace of Dom Pedro II, with its luxurious furnishings, then returned to Rio, where we bought some aquamarine gemstones before heading home.

New Directions

In Cambridge, I came back down to earth with a thud, gearing up for classes at MIT and a full consulting schedule. All in all, it was a hectic but productive period. A year or so later, in a campus-wide effort to gauge faculty productivity and in a departmental effort to determine how much the Department of Electrical Engineering should encourage outside consulting, Professor Hazen asked all faculty members how we spent our time each week. I estimated that, on average, I was spending 15 hours teaching, 12 hours with students and my own research in the Acoustics Laboratory, 20 hours in outside consulting, 5 hours in book writing, and 8 hours in uncompensated public activities. Hazen was not pleased. "Unless at least a portion of one's fruitful time," he told me, "is left for calm and unhurried reflection, the longer-term goals are likely to suffer loss to the short-term goals." Although I found it hard to change the habits I'd acquired in the course of my wartime work, I did cut back a little.

In 1952, Gordon Brown became head of the Department of Electrical Engineering. His first goal was to transform the curriculum from its emphasis on electrical machinery to the realm of modern electronics. Thirty years later, he recalled his thinking in *A Century of Electrical Engineering and Computer Science at MIT* (MIT Press, 1985). "I made my first approach to getting a little emancipation into the electrical machinery area by inviting Leo Beranek, then one of the key people in the Acoustics Laboratory, to begin quietly collaborating with me on an idea that struck me as pertinent, namely, that a loudspeaker is in principle the same as a motor. It was certainly an energy conversion device, with a dynamic performance far in excess of any rotating machine; nonetheless, it could be described by the same equations. I thought that Leo could be of great help in revising the machinery curricula." But the path, he observed, was riddled with obstacles: "I had to learn that you approach such changes very discreetly in a place like MIT. The acoustics people objected because they thought that Leo was selling himself out to the machinery people, and they thought I was intruding in their areas. So we didn't move very fast." Brown went on to take steps all his own leading to the adoption of completely new courses under young faculty members. But, even so, as a member of the teaching force for two of those transformative courses, in two successive years, I was part of the revolution.

At this time, my primary MIT-based research focused on the noise generated by heating, ventilating, and air-conditioning (HVAC) systems. I interested first two, then two more, Navy and Coast Guard students in working with me on measuring noise generated by fans in HVAC systems. These studies involved the purchase of several types of fan, and installing them in such a way that we could measure noise at low, medium, and high frequencies for different fan speeds. The results, published in the *Journal of the Acoustical Society of America*, became the basis for HVAC standards worldwide.

The Acoustical Society of America

One of the greatest professional influences on my life, the Acoustical Society of America with a current membership of about 7,000, enables those in the field to share new technical findings at semiannual meetings and through its journal (*JASA*). I remember one of the first meetings I attended, in Iowa City, fall of 1939. Professor Ted Hunt presented the work I had helped him with and I got a good feel for how new research is communicated to colleagues.

One day after the noon break between sessions, while walking up a hill near the University of Iowa, I was overtaken by Dean Carl Seashore, famous for his studies of musical aptitude. He asked about my research and gave me advice on how to bring my findings to the attention of our colleague. I was agog that such an icon in the field would go to this trouble.

The papers I have presented at meetings and published in *JASA* have built personal and professional relationships that have meant more to me than I can possibly express here. They also brought me my earliest professional recognition, starting with the R. Bruce Lindsay Award, given to a member of the society under 35 years of age, which I received in the spring of 1944. The award helped cement my affiliation with the society: I was elected a few months later to a three-year term on the Executive Council.

As a council member, I had my first opportunity to tackle some touchy factional issues. Because they were denied greater autonomy under our umbrella by our secretary, Wallace Waterfall—who had not advised the Council of the audio engineers' request—an important group of our members had just broken off to form the Audio Engineering Society. Waterfall and our treasurer, Herbert Erf, both of whom were, in effect, automatically reelected from year to year, had grown accustomed to making nearly all the decisions and involving the Executive Council very little, if at all. I let it be known in no uncertain terms that I had joined the council, not as a ceremonial duty to rubber-stamp executive decisions, but to deliberate policy in a serious, responsible way.

The society as a whole seemed to appreciate my stand: I was elected president in the summer of 1954. That fall in Austin, Texas, at the general meeting, open to all members, I asked the gathering for comments on the society's future. A number of members stepped forward to air pent-up feelings that, if held back, could have resulted in further splits in the membership. In response, I created the Committee on the Development and Promotion of the Society, with R. Bruce Lindsay as chairman, to recommend ways to defuse a crisis waiting to explode. The committee recommended establishing "technical sections" for the society's subspecialties—building acoustics, noise control, psychoacoustics, underwater sound, and so on. With the Executive Council's approval, ten technical sections were established within two years. Each section would meet for about three hours one evening during the society's semiannual meetings, to mull over and vote on questions unique to its particular subspecialty.

In 1957, now as chairman of the Promotion and Development Committee, I met with the chairmen of the technical sessions. After getting an earful from them, I was granted permission to report the chairmen's recommendations directly to the Executive Council. After due consideration, the Executive Council decided to form a separate Technical Council comprising the to-be-elected chairmen of the ten technical sections under the cochairmanship of the society's vice president and vice president–elect, who had power to ensure that the needs of the technical sections would be considered by the Executive Council. Thanks, at least in part, to the more balanced distribution of executive control, there were no more splits in the society's membership.

Housing and Home Life

My life outside work was going well, too. Phyllis had friends in Boston, where she now worked as a dental hygienist, and we got together with them quite often. A year and a half after Jamie was born, we decided to look for a suburban home.

In 1950, having selected a 1-acre parcel in a new development in Lexington, we were about to engage a builder when an alternative surfaced. We were visiting with Thomas and Elizabeth Ballantine at their Boston home (Tom was a leading neurosurgeon at the Massachusetts General Hospital; Elizabeth, whose family dated back to colonial America, was a long-time friend of Phyllis). Hearing about our building plans, Elizabeth told us that a dear friend of hers in Winchester, Mary Parker, whose husband had died recently and who could not bear to continue living in the house they had built just two years before, had decided to sell it. If we were interested, she would ask Mary to show us the house.

Mrs. Parker told us her lawyer had urged her not to sell immediately, but to rent out for two years and then decide. The place was ideal for us. Set on two wooded acres, fully landscaped, the two-story house boasted three bedrooms—ample for all our current needs—and stood on the highest point in Winchester. When Mrs. Parker offered to rent to us for $200 a month, we accepted. After selling our Lexington parcel to one of the development's architects, we moved in a few days after Christmas 1950.

By now, Phyllis had quit her Boston job. She would take Jamie for long walks in Cambridge. With babysitters to tend him, we would go out at night, often to concerts and plays. Our Winchester house had a playroom adjacent to the kitchen, where Jamie could play with his toys. Our dog and two cats were a source of constant entertainment for him. Jamie also had four neighbor children his age to play with; they would wear themselves out racing round and round with their tricycles in our large circular driveway. Each evening at the dining room table, I would read aloud from books—choosing stories to match Jamie's growing curiosity and comprehension.

A nanny came to the house one day a week, freeing Phyllis to get away. On that evening, we usually went into Boston for dinner. Each January or February, we would ski in Switzerland for a month, with the nanny taking care of Jamie while we were away until he was old enough to come along. When Jamie was 9, we decided to have a second child. We named our new son Tom while reading Mark Twain's *Tom Sawyer* at the dinner table. Jamie attended public school in Winchester through the seventh grade and Saint Mark's School from the eighth through the twelfth. Tom went to Belmont Day School through sixth grade and Fenn School for seventh and eighth grades, following his brother to Saint Mark's for high school.

Teaching and Consulting

At work, meanwhile, requests kept coming in to help architects design acoustically friendly office spaces. By now, I was juggling two different roles—one in academe, the other in business, the consulting firm Bolt Beranek & Newman (BBN, whose origins are outlined in chapter 5)—but with a marked overlap in mission, practice, and research content. Having no idea how quiet the ideal office should be, I set out to first determine what noise levels were acceptable. With student help, I surveyed ambient sound—and responses to it—in different offices at MIT, General Radio, and a metal fabricating company. As we took our noise measurements, we would ask people working in that space to assess the noise level. They were given a rating sheet and asked to (1) rate the noise level on a scale from "very quiet" to "intolerably noisy," with intermediate steps of "quiet," "moderately noisy," "noisy," and "very noisy"; (2) rate their ability to use the telephone at the prevailing noise level on a scale from "satisfactory" to "impossible," with intermediate steps of "slightly difficult" and "difficult"; (3) select a maximum noise level from the scale used in (1) at which they could accomplish their duties without loss of performance; (4) state the maximum distance at which they should be able to converse with a coworker and still accomplish their duties without loss of performance; and (5) rate their own typical speech level—"normal voice," "slightly raised voice," or "loud voice." With this as a start, we applied for and were awarded a government contract to continue the study in depth at an Air Force base in the Rocky Mountain states. Our results made it possible for the first time to state how employees in different types of space rate the acceptability of a noise at low, middle, and high frequencies. The "noise criteria curves" we developed became the basis for national and international standards and are used to this day.

As the years went on, my work at Bolt Beranek & Newman ate up more and more of my time. I reduced my commitment at MIT to 75 percent in 1951. In the spring of 1953, I was elected president of the Acoustical Society of America, with a one-year training period as president-elect, before formally taking office. This new position, along with my escalating duties at BBN, so diminished my efforts at the Acoustics Laboratory that I felt I could no longer serve as an effective technical director. After careful consideration, in the summer of 1953, I went to Gordon Brown and asked him to reduce my time still further, to 50 percent, to relieve me of my duties as technical director, and to arrange for me to move my office and research space to the main MIT building under the Great Dome, where I would not be taking up valuable space in the Acoustics Laboratory. I told all this to Dick Bolt, too, who understood; after I moved, he reassigned my office to his administrative assistant, John Kessler.

In 1951, the courts in New York State ruled that employers would be held responsible for progressive loss of hearing incurred over a period of years from work in high-noise environments. This decision led to a "damage to hearing" scare. Even though, by then, I was easing out of MIT commitments and not eager to add new ones, I went to Gordon Brown in the summer of 1952 and recommended that a summer course be established at MIT on noise and noise control, and that it be geared toward both industry and academe. Gordon thought this a fine idea, told me to work with the head of Summer School Programs to develop such a course, and advised me to select as many of my instructors as possible from MIT, although he gave me permission to use some personnel from Bolt Beranek & Newman as well.

So it came to be that MIT offered a special two-week course on noise reduction in August 1953 under my direction. Enrolled were 120 people from industry, government, and academic institutions. The course was taught again in 1955, 1957, 1960, and 1964, with enrollments of 165, 185, 140, and 110, respectively—the largest attendances for a single summer course at MIT to that time. The lecturers were drawn from the MIT faculty and Bolt Beranek & Newman—I gave about a third of the lectures. At the end of the course in 1964, however, the head of the Summer School told me he was closing it down because it had become too much of a Bolt Beranek & Newman affair—more than half the instructors were from BBN. I was not sorry about this: my responsibilities at BBN continued to increase and I had many irons in the fire.

Meanwhile, the Acoustics Laboratory flourished into the mid-1950s, during which time its principal support came from the U.S. Navy, Bureau of Ships. The lab successfully developed an important acoustical material for use by the Navy. Visco-elastic, thin, and sound-absorbent at sonar frequencies over a wide range of temperatures and ambient pressures, this material appeared to be the ideal coating for submarines, which would render them nearly invisible to enemy sonars—a critical development as the Cold War gathered momentum.

The Bureau of Ships was so taken with the prospects that they authorized coating an entire submarine for testing at sea. In retrospect, the laboratory should have simulated the test beforehand, perhaps using some sort of box made of the same steel and coated in the same manner as the actual test submarine. The box could have been placed in the ocean to test its durability. But it did not. Instead, barnacles and other encrustations were removed from the surface of the test submarine with grinding equipment at the Boston Navy Yard. It was at this point a major difficulty arose. After the surface had been cleaned but before the visco-elastic coating was applied, enough rust formed to interfere with bonding. Failing to anticipate what might happen, shipyard personnel went ahead and cemented the coating in place. When the vessel hit its first stretch of rough water, however, half the coating peeled off. Bureau of Ships officials were so angry that they terminated their contract with the Acoustics Laboratory (unjustifiably, in my view). The laboratory closed on January 31, 1958. With no financial support, Dick Bolt resigned his MIT professorship and went to Washington, D.C., to serve as associate director of the National Science Foundation.

Research and Industry Ties: The Bell System

In 1945, while still at Harvard, I had been approached by Harry C. Tuttle, president of the Hush-A-Phone Corporation in New York, to develop a new privacy device—a "scoop" attached to the handset that, when talked into, would provide privacy and reduce ambient noise. The company had been merchandising an early form of the device for over two decades, but it could not be used with AT&T's newest handsets because of its shape. I completed my design in 1946 and filed for a patent, which was awarded two years later.

In lieu of a fee, I negotiated a license agreement with Hush-A-Phone for a royalty of 20 cents per unit and a minimum of $500 a year. My design did a much better job than the existing one of enhancing privacy and eliminating external noise. The new device was made of black Bakelite, about the size of a fist holding a baseball, and, when in place on the handset, fitted comfortably around the face. The Hush-A-Phone Corporation sold it through advertisements in the *New York Times* and other prominent news outlets, the principal buyers being bankers, financial advisers, and secretaries in medical offices.

The corporation had discovered, however, that Bell Telephone servicemen were warning Hush-A-Phone users—by reading them a boilerplate statement prepared by Bell—that they were in violation of FCC rules and that if they continued to use the device, they risked having their telephone service terminated. AT&T justified this tactic, which was clearly aimed at scaring telephone subscribers away from Hush-A-Phone, on the basis of Federal Communication Commission tariff 132, which read: "No equipment, apparatus, circuit or device not furnished by the telephone company shall be attached to or connected with the facilities furnished by the telephone company, whether physically, by induction or otherwise." On December 22, 1948, Hush-A-Phone filed a complaint with the FCC, requesting that it amend Tariff 132 to permit the use of Hush-A-Phones, and that it instruct AT&T to cease and desist its campaign against Hush-A-Phone's distribution and use.

The FCC held a public hearing in Washington, D.C., in January 1950. The cavernous hearing room was intimidating in its own right with a raised bench on one side for the judges. Both the FCC and AT&T recognized that this was an important, possibly a precedent-setting case. Three examiners were assigned to hear it—Nicholas Johnson from the FCC; Harmon Fowler and Walter F. Roberts from other federal commissions. Harry Tuttle engaged Kelley Griffith, a young attorney with little trial experience, to argue on Hush-A-Phone's behalf. I was the only expert witness for our side. Representing the Bell System, on the other hand, were dozens of attorneys, a number of expert witnesses, and a top trial lawyer imported from New York City. The defending parties included not only AT&T but 21 regional telephone companies as well, all part of the Bell Telephone System. To seat them all, folding stepped "bleachers" had to be brought into the room. Incidentally, not one of AT&T's expert witnesses was from its highly respected research arm, the Bell Telephone Laboratory—all came from the telephone system.

In preparation for the hearing, I had Tuttle hire a colleague of mine at MIT, Professor J. C. R. Licklider, to conduct articulation tests demonstrating that our device did not impair speech intelligibility. I assembled charts to show how the attachment worked and to illustrate Licklider's findings. Tuttle, called as the first witness, showed our device to the court and described its sale and where it was used. AT&T's lawyer, in a tough cross-examination, tried to make Hush-A-Phone look small potatoes as a company. I then took the stand to explain technical aspects of the design and our various tests. When AT&T's expert witnesses were called, it was clear that they had all been well versed in a test purporting to prove Hush-A-Phone's negative effect on speech intelligibility. But the test, conducted some twenty years earlier, was irrelevant to our new device's performance; their "expert" testimony was effectively refuted by Licklider's unimpeachable results. I pointed this out and went on to explain that Hush-A-Phone offered about the same privacy as the common public, glass-enclosed telephone booth.

On cross, AT&T's attorney had me describe in eye-glazing detail how each of Licklider's tests and my privacy tests had been carried out. Getting nowhere in undermining the integrity of those, he switched tactics, trying to make me look foolish, as he had tried to do with Tuttle. When I told the court how our device reduced surrounding noise, and that I used it in my office for that very purpose, the attorney asked if I knew that AT&T could supply a handset with a button on it that, when pressed, would disconnect the transmitter and do essentially the same thing. Yes, I said, but also noted that AT&T charged a hefty monthly rental fee for this service. "Dr. Beranek," he rejoined sarcastically, "I understand that you are an important professor at MIT, and for someone of your importance MIT should be glad to give you what you want regardless of cost." I shrugged off this remark, pleased to imagine that I had unnerved, however slightly, this brilliant, high-paid corporate attorney.

Next, the court permitted Tuttle to demonstrate our device. He had someone telephone to the hearing room and, while talking, take the Hush-A-Phone on and off the handset to show its impact on speech intelligibility. Although there was a significant change in voice quality when the device was attached—a "boominess" was added—every word could be understood without difficulty. AT&T's attorney countered that voice quality was as important as speech intelligibility, and that this demonstrated "harm" to the telephone system.

At one point, one of the judges asked an AT&T witness whether using the Hush-A-Phone had the same effect as cupping a hand around the mouthpiece while talking. The witness agreed that the effect was indeed similar, observing that our device was therefore not merely an encroachment but also a superfluous contraption. The trial dragged on for over two weeks. I came back to Cambridge exhausted, after having been on the stand for 12 hours on each of two days. The FCC rendered its final decision on December 21, 1955—an incredible five years later—ruling that tariff 132 should be retained without alteration.

The Hush-A-Phone Corporation next took the case to the U.S. Court of Appeals, District of Columbia Circuit. In November 1956, this three-judge panel ruled in our favor and rebuked the FCC for its five-year lapse in reaching a decision. The court ruled that tariff 132 was an "unwarranted interference with the telephone subscriber's right reasonably to use his telephone in ways which are privately beneficial without being publicly detrimental." The ruling further stated: "To say that a telephone subscriber may produce the result in question by cupping his hand when speaking into it, but may not do so by using a device which leaves his hand free to write or do whatever else he wishes, is neither just nor reasonable." And, in regard to the change in sound quality: "Indeed the cupped hand may distort [the words] more than the Hush-A-Phone, for the Hush-A-Phone is provided with an acoustical filter and ducts which partially absorb the low frequencies; holes are also provided in the Hush-A-Phone through which the low frequencies are partially conveyed to the outside. The air blast effect is also reduced by releasing the air through the holes." The court based much of its analysis, and even some of its wording, directly on my testimony.

Hush-A-Phone's legal victory opened the way for yet another customer-owned device. Invented by Thomas Carter, the Carterphone was a specially designed base on which a telephone handset could be placed. In this base was a microphone to pick up whatever was coming into the receiver end of the handset and a small loudspeaker to send messages into the transmitter end. Carter conceived of this device as a way to connect mobile radio and ship-to-shore communication networks directly to the telephone system, this circumventing similar services offered by AT&T at a far higher cost. AT&T cited FCC tariff 132 once again, hoping to shut out Carterphone. In 1966, Carter sued AT&T for the right to connect the device and the case was referred to the FCC. This time, the commission used the Hush-A-Phone case as precedent for repealing tariff 132 altogether, and on June 26, 1968, handed down its decision that established the right of telephone subscribers to connect their own equipment to the public phone system, as long as any such equipment caused no harm to the network.

Essentially, the floodgates were now open. Customers could connect foreign devices of all kinds—fancy telephone sets, faxes, modems, alarms, answering machines, even local private telephone exchanges—to their telephone lines. It is no exaggeration to suggest that, without the ability to connect modems to the network in this way, the Internet would not have become the global operation that it is today.

In 1971, in yet another setback for AT&T, the MCI Corporation petitioned to establish a telephone network from Chicago to Saint Louis. The FCC certified MCI as a "specialized common carrier," citing, in particular, its data transmission services, which were far superior to those offered by AT&T. Again with the Hush-A-Phone decision as precedent, MCI was allowed to connect to the Bell System but could not charge for long-distance services in areas served by AT&T. As it turned out, MCI did not cut much into AT&T's income, which increased rapidly as the demand for long-distance service began to soar in the 1970s. AT&T's size and political power grew to such gigantic proportions, in fact, that the U.S. Justice Department brought antitrust action against the company in 1974. MCI filed its own suit simultaneously, seeking the right to sell services to Bell's subscribers. Various providers of telephone equipment—eager to sell equipment directly to the Bell System—also threatened court action. In 1984, faced with the prospect of multiple lawsuits and some not-so-gentle prodding from Congress, AT&T entered into an antitrust settlement that divided the Bell System into seven regional Baby Bells. The settlement also required that every Baby Bell purchase products and services from a competitive field of suppliers, instead of exclusively from AT&T and its manufacturing arm, Western Electric. Hush-A-Phone may be all but forgotten amid the spectacular array of services available to us nowadays, but the record is clear nonetheless—it marked the start of a process that forever changed the face of telecommunications as we once knew it.

Work, Play, and a Mix of Each

While at MIT, I wrote four books: *Acoustics* (McGraw Hill, 1954), adapted from courses I taught in the Electrical Engineering Department; *Noise Reduction* (McGraw Hill, 1960), derived from the special summer programs at MIT;

Music, Acoustics and Architecture (Wiley, 1962); and *Noise and Vibration Control* (McGraw Hill, 1971)—the last two done jointly with staff from Bolt Beranek & Newman.

Needing some diversion from my work at MIT, I found one I truly enjoyed—skiing. In the late spring of 1949, I paid a visit to Professor Smitty Stevens, with whom I had worked closely at Harvard during the war. Smitty sat in his spacious office behind a piled-high desk and scowled at me. Bushy eyebrows, an athletic figure though a little hunched, this brilliant experimental psychologist could be a formidable—some might even say overbearing—presence, but to me he was simply a good friend. I was there to ask him to take part in a seminar at MIT, and he accepted. These formalities over, he jumped quickly to his new hobby. He had purchased a house in North Conway, New Hampshire, near the Mount Cranmore ski resort, and learned to ski from a German, Otto somebody. He urged me to take up the sport and to get Phyllis into it as well. He invited us to spend weekends with him and to take lessons from Otto, assuring me that now was an excellent time to buy skis: the sporting goods store he went to on Bow Street in Cambridge had everything at half off. I talked this over with Phyllis and we agreed it was worth a try, so off we went to Bow Street.

Our first year of skiing was a series of difficult learning experiences, with aches and pains each night, but we enjoyed our evenings with Smitty and his companion, Didi. We went shopping with them and helped prepare meals. Mostly, we discussed which slopes were best for beginners and how to improve our technique. The second year we took our son, Jamie, who was three years old and learned more rapidly than we old fogies. The weather did not always cooperate. The first ample snow usually came in January, but the temperature always fluctuated above and below freezing. There was enough in the way of rain that I invested in a rubber suit with a rubber hood, and some days found myself skiing almost alone—few souls, it seemed, were hardy or crazy enough to brave the uncomfortable wet conditions. One weekend, the slope was so icy that I fell and injured a rib. We never skied in New England again.

The following January, 1953, the three of us ventured out to Alta, Utah, home of the finest powder anywhere. I took private lessons and improved a great deal. A year later, I was attending an international meeting in Bern, Switzerland, when Willi Furrer, a senior officer in the Swiss telephone company, invited me to go skiing with him in Murren, a village in the Bernese Oberland. With a view of the Eiger, Jungfrau, and Munch, Murren has to be the most beautiful place to ski in all of Switzerland. I did not ski as well as Willi, but we had fun. By the next winter, his secretary had found us a rental house in Murren—Chalet Flueblumli—where Phyllis, Jamie, and I vacationed for several years afterward.

In January 1956, Phyllis being pregnant at the time, Jamie and I went to Alta together for three weeks. I took lessons from the head of the ski school, Alf Engen, to gain proficiency in deep-snow skiing. The following year, after Tom was born, we went to the charming Swiss village of Grindewald. I saw a sign—"Test your ski capabilities by means of the Swiss Ski Tests"—and accepted the challenge. I decided on the "Silver Badge Test," which, according to the literature provided, "proves that you are an efficient skiing tourist capable of undertaking all tours without being a hindrance to the party." The day before the test, I engaged a ski instructor in the hope that he would "fine-tune" me. After an hour together on the slopes, he said I should aim higher—the "Gold Badge Test."

Two days later, I showed up and found that I was the only candidate for the Gold Badge. The examiner and I took a train to the Jungfrau getting off halfway to the top, right where the train disappears into the mountain, at the base of the Eiger. We traversed laterally on skis for about ten minutes until we came to a spectacular slope, a quarter mile long, very steep, and over 500 yards wide. About a foot and a half of new snow had fallen the night before and no one had yet sullied the smooth surface, which glistened in the morning sunlight. The examiner turned to me and snapped, "Ski it!" Having never felt better or more confident, I skied the slope, almost to perfection, sending billowing plumes in first one direction and then the other. When I pulled up at the bottom, the examiner stopped behind me and said: "There are not more than a dozen skiers in Switzerland who can handle deep snow better than you." The course I had just run, along with his glowing assessment, left me feeling exhilarated.

Other test requirements included following the examiner in and out of difficult spots and making U-turns on steep slopes. But the most daunting hurdle of all was a timed run down the Lauberhorn. The examiner and I took the lift to the top. He unfolded a large red flag from his rucksack and announced that he would make a run down the mountain first, to establish the time against which I would be judged. At the bottom, he would wave the red flag signaling for me to start and he would time my run, I had jumped through all the hoops of this examination and was not about to muff this one. I tightened my safety bindings. When the flag dropped, I took off and skied like a wild man, everywhere just barely short of losing control. When I reached him, he declared I had passed. The standard for the "Gold Badge" test reads: "approximately corresponds in actual ski-running to the standard required for the [Swiss] ski-instructors' examination. It is the badge of a first-class skier."

In succeeding years, Saint Moritz became our best-loved spot for skiing, with several mountains available, the choice on any particular day depending on the weather. We reached the main peak, Piz Nair, which rises to a little over 10,000 feet (3,055 meters), by a basket lift. A twenty-minute drive from Saint Moritz are the twin resorts of Diavolezza and Lagalb. Diavolezza had beautiful trails, but they were usually crowded. We tended to go there on days when we could expect the fewest day skiers. Sometimes we would race down the Diavolezza-Morteratsch Glacier. This was a scary trip, with paths ranged above steep slopes—a slip might cost a life. But the biggest challenge for my ski partner Tom and me was Lagalb, where a large official sign next to the parking lot proclaimed: "Lagalb, for pretentious skiers." Just one cable car ran to the pinnacle, and only two distinct trails ran down. You could ski at almost limitless speeds; because only daredevils attempted them, the slopes were quite empty. We found we could reach the bottom in fifteen minutes, just in time to catch the cable car on its way back up.

The International Circuit—Professional Connections

Back on the work front, in July 1967 I received a letter from L. Clarkson, professor of Vibration Studies, offering me the directorship of the Institute of Sound and Vibration Research in Southampton, England, which, the professor believed, would "strengthen our links with work in the USA and particularly with that of Bolt Beranek and Newman." I wired my regrets. I did not say so, but the truth of the matter was that I did not care to live in England and my salary would have been half of what I was getting from my consulting work.

On the other hand, I always got a great deal out of my professional trips overseas. In the spring of 1957, I was invited by the Institution of Mechanical Engineers in London to deliver their Forty-Fifth Thomas Hawksley Lecture, scheduled for 1959. I was free to speak on any subject that might interest the membership. It was an unusual invitation—only one other American had ever been the recipient of this high honor. I worked for nearly two years on "The Transmission and Radiation of Acoustic Waves in Structures," a topic in the vanguard of acoustical studies at the time, and produced what still stands as one of the most complete treatises on the subject. But the lecture itself did not generate much satisfaction with some who spoke to me afterward saying it went "over my head." Apparently, the members were accustomed to being entertained rather than educated at such events. I agreed to repeat the lecture in Oxford and Bristol, and, with the London experience behind me, made my presentation considerably less formal and more anecdotal.

Having been invited several times to speak at the Technical University in Zurich, the University of Warsaw, and the Institute of Acoustics in Moscow, I wrote to all three, asking if they would now be interested in hearing me on the topic of my London lecture. Each responded enthusiastically, and the Moscow people invited me to stay two weeks and to speak at several venues. Because my Beranek forebears had emigrated from Prague in the mid-nineteenth century, I was curious to go there as well. I wrote to Professor Slavik of the University of Prague, saying I would be flying from Switzerland to Poland and my family and I would like to stop over in Prague along the way. An invitation to lecture at the university arrived forthwith.

Phyllis, 11-year-old Jamie, and I thus embarked on a trip that was, and still is, the longest I have been out of the United States, except for Buenos Aires. Tom, a mere toddler, stayed home with a nanny. I gave the lecture in London on November 21, 1958, and we didn't return until mid-January 1959—nearly two months abroad. Local expenses were covered by our hosts in each city, except Moscow, although there they reimbursed us for our round-trip air fare from New York to Moscow—and in American dollars. I agreed to give three lectures in Moscow—the one I gave in London, a second on concert halls, and a third on noise around airports.

Our most striking experiences were in the Communist countries. We had a chauffeured car in Poland and Czechoslovakia, always with a professor in the front seat when we were taken sightseeing. We suspected the driver had a recording apparatus, presumably to keep our tour guide on the straight and narrow. Actually, the degree of caution differed considerably between the two countries. In Czechoslovakia, people talked personally only when walking down the street with no one nearby. In Poland, on the other hand, people joked freely about the government and, in particular, about the dour Kremlin-dictated architecture of their principal buildings. Professor Maleski of Warsaw University told me that he had talked to Moscow about our upcoming visit there and had learned that I was to receive royalties on my book *Acoustic Measurements,* which, unbeknownst to me, had been translated and published there. Incidentally, customs officials in Moscow gave me trouble with both the antique pistol and the coins I bought in Prague at very low prices.

We flew on the Soviet airline, Aeroflot, from Warsaw to Moscow. Safety seemed the least of their concerns: there were no seat belts and heavy packages were placed, untied, on empty seats. The flight attendants handed out crude sandwiches and tea. We arrived in Moscow after dark and were greeted at the airport by Academician N. N. Andreev and the head of the Acoustics Institute, Leonid Brekhovskikh, as well as several other acousticians. A pair of limousines whisked us to the Hotel Metropol, Moscow's finest at the time. Phyllis and Jamie rose in one car and I, with the "big wigs," in the other. They told me that my lecture on concert halls would be very well attended, but that just a small group of specialists would hear me talk about airport noise. Yet curiously, when I gave the lectures, just the opposite happened. Because the authorities were planning a new concert hall they apparently did not want my speech to conflict in any way, so they limited the attendance at the last minute and, to compensate, encouraged more people to attend the airport noise lecture.

What an eerie feeling to arrive in the Soviet Union, to pass by the Kremlin with its huge internally illuminated red stars rotating on the tops of several towers, to gaze wide-eyed as we headed through Red Square, and then to end up at the Metropol. Brekhovskikh gave me a few hundred rubles, saying I could repay the money when I received my royalties. We were escorted to a grand suite with three rooms, one of which was a living room with a library and grand piano. Jamie had his own bedroom.

On our first day, I went over to the Acoustics Institute and met a number of the staff while Phyllis and Jamie were taken to art galleries. The second day, I was told that a car would arrive at 4 PM and take me to the publishing house to collect my royalties. At the appointed time, an interpreter hailed me in the lobby and said that I should bring a "suitcase" to carry the rubles. I brought a small travel bag. At the publishing house, a pile of cash equivalent to $15,000 in ten-dollar denominations was placed in front of me. The officials there insisted I count it, and going one by one through 1,500 bills proved no easy task. When we got back, I said I did not want to have so much money with me in the hotel, so the interpreter took me to the Bank of Moscow, a huge stone building, which was closed when we got there. By standing on a ledge outside, he could reach a window, which he pounded on until someone inside yelled at him through the glass. He shouted back my request, and soon the great front door moved to one side and we were waved through (so much for the stereotype of rigid Soviet bureaucracy). Along one side of the lobby was a row of teller windows, and one of those lit up. I went over and deposited about $10,000, in exchange for which I received a bank book. I paid my expenses in the Soviet Union out of these royalties.

On about the fourth day, I was invited to tour Moscow's phonograph recording house, although I had expressed no particular interest (Phyllis and Jamie chose not to go). Picked up the next morning about 10 AM, after a formal meeting in the director's office, I was taken to Studio 1. Through two sets of double doors (sound traps), I came upon a studio the size of a small auditorium, but with no permanent seating. Just inside the second set of doors stood a short, rotund man about 50 years old, holding a violin, whom my staff escort introduced as David Oistrakh. "Mr. Oistrakh," I stammered, awestruck, "in America, we consider you the world's greatest violinist." He smiled and thanked me through an interpreter.

At the far side of the studio was a large stage and on it, to my amazement, sat the Moscow Philharmonic Orchestra. The only seats for listeners were four upholstered chairs in the middle of the floor. I was escorted to this area, along with three officials from the recording house. As I sat there flabbergasted, David Oistrakh and the Moscow Philharmonic performed Beethoven's Violin Concerto just for me. I kept thinking there must be some hidden agenda here—and so there was. At the finish, I was told that we would go to a playback studio to listen to recordings of the performance that had been made using two different configurations of microphones above the orchestra. There had been an ongoing argument, apparently, between the recording engineers and Oistrakh about the best placement of microphones. In the playback room I was asked to judge the better of the two. The engineers played back ten minutes of what they called "A" and then the same segment of "B" (different microphone locations). I heard a distinct difference. I asked them to play "A" once more and immediately stated my preference for "B." Oistrakh jumped up, strode over, and gave me a big hug. I didn't look in the engineers' direction, but I assume they were a bit chagrined. A week later, I attended a performance by Oistrakh in Leningrad (now called "Saint Petersburg" once again) and had my picture taken with him. When we left Moscow to head home, we ran into him again—this time by chance—at the airport, on his way to another international tour.

To get to Leningrad, we took the overnight sleeper train. Each bedroom accommodated two, and for a single ticket a passenger might be paired with anyone. On the way there, I roomed with a Soviet general; on the way back, Phyllis roomed with a Soviet woman. Using our large stock of rubles, we bought antiques, a movie camera, and several presents for Jamie. Our last night, we hosted a sumptuous banquet in the Metropol for the acousticians whom I had come to know, and their spouses. Our guests, I learned, had never before experienced an event like this, where both husbands and wives were included. We spent Christmas in Vienna and New Year's in Tel Aviv, then we skied for two weeks in the Swiss resort Klosters.

In 1958, I resigned my tenured faculty position at MIT to work full-time at Bolt Beranek & Newman. This greatly upset Gordon Brown, who urged me not to go, offering to reduce (or even eliminate) my teaching schedule and to be content if I showed up only at faculty and special meetings. Thanking him for his offer, I said it would be unfair for me to take the place of someone who could give a full-time commitment to the department. As noted earlier, however, I would remain active as a lecturer in MIT's summer courses for a number of years.

5 Bolt Beranek & Newman, the United Nations, Big Noise, and the Internet

The wave of postwar optimism still ran high in 1948; Cold War clouds were gathering, but would not thicken and lower for a few years yet. The MIT campus teemed with activity. We had a new crop of graduate students, a new president (James Killian), and a new Acoustics Laboratory. I had been around just a year, but already felt quite at home and positive about the trajectory of my work.

Although large, my office also housed my secretary in a cozier arrangement than some at Harvard might have tolerated. Students dropped in regularly, and if we had confidential matters to discuss, we would move to a nearby conference room. The lab—indeed MIT as a whole—didn't stand much on ceremony, and its casual, nonhierarchical atmosphere helped foster friendly give-and-take among faculty, support staff, and graduate students. I remember the look of shock on the faces of foreign visitors when they heard my secretary call me "Leo." A professor from Delft told me that he could not countenance having his first name (which happened to be Kees) appear anywhere in a letter, for fear his secretary might start using it.

This aura of social informality served to offset the rigor and intensity we all brought to our work. MIT demanded commitment, focus, dedication— no fooling around when it came to our teaching or research. As scientists engrossed in our work, we worried little about physical comforts. We simply disregarded minor irritations and pressed ahead. The pressure of work was such that fatigue overcame me some afternoons. When I found myself nodding off while talking to a student, however, I discussed this with my doctor. He suggested amphetamine tablets, but I worried about side effects. We settled on a mild sleeping pill, which gave me a longer night's rest and my second wind in the afternoon.

Dick Bolt was director of the laboratory and I the technical director; we had equal status in the handling of technical matters, but administrative and financial matters fell entirely to him. As part of the postwar building boom, requests kept flowing in to MIT from architects and contractors around the country seeking help with acoustics problems. Whenever a request would come in, the president's office routinely called Dick, and he and a graduate student would give advice, which generated side income for Dick.

President Karl Compton had made it clear when he hired me that I would be allowed some time for outside activities. MIT's policy allowed faculty members one day a week for consulting, plus weekends and summers. One professor told me that it was not unusual for someone to equal—in rare cases, even exceed—his academic salary in outside fees. The MIT attitude was quite different from what I had found at Harvard, where colleagues tended to look down on paid outside work as undignified at best, and as moneygrubbing at worst.

The 1948 fall term at MIT was more intense than the previous year. I opted to be the junior teacher in the Electrical Engineering Department's electromagnetics course, even though I had never studied the subject, which meant extensive preparation, to keep one step ahead of the class. This was on top of preparation for my acoustics course; though I knew the material backward and forward, each lecture still had to be carefully honed because I always spoke without script.

Dick's outside activities mostly involved architectural acoustics projects. I had three outside consulting jobs. The most challenging was at the Cambridge facility of General Radio Company, where I spent every Friday afternoon helping to develop a new line of acoustical measuring equipment. I devoted some time each month to the Hush-A-Phone Corporation project described in chapter 4. My third job was an exciting new venture upgrading the acoustics of fifty movie theaters in Brooklyn, New York.

Partners in Business

A big surprise came in October 1948. Asking me to come across the hall to his office, Dick pointed to a pile of drawings, 8 inches thick and 10 feet long, plans for what would become the permanent seat of the United Nations in New York City. He had originally heard about the project through the office of MIT's new president, Jim Killian, who had just succeeded Karl Compton.

The New York architectural firm of Harrison and Abramowitz was asking for bids on a major acoustical project. Dick had gone to New York to learn more about the project and had prepared a responsive bid. He later learned that Vern Knudsen of UCLA, himself a distinguished acoustician, had submitted a much higher bid because of the extra time and expense involved in commuting from the West Coast. It was not until the mountain of drawings was dumped in his office that Dick realized the magnitude of what he had taken on. The project was too much for one person, and he wondered whether I would work with him on it as his partner. I agreed on the proviso that we enter into a full-fledged partnership and that all of our consulting work, from then on, be done through it. Dick thought this a great idea.

We contacted a lawyer, whose first question was: what will you call your new firm? "Beranek and Bolt," I piped up, addressing myself to Dick. After all, my name preceded his alphabetically and I was well known nationwide as a result of having directed two Harvard laboratories. Dick balked, however, proposing instead that his name come first because he was the one who had gotten the United Nations contract and come up with the partnership idea—plus, he was senior in both age and position at MIT. I thought a few seconds and agreed. "Let's go with 'Bolt and Beranek,'" I said, "I want this to be a happy collaboration." The firm was formally inaugurated on November 8, 1948.

Our new partnership had Jim Killian's blessing. He offered to rent us two rooms in the Acoustics Laboratory, with the understanding that when we outgrew that space we would locate elsewhere. We were allowed to use MIT's address of record, 77 Massachusetts Avenue, although without any reference to MIT. We occupied this space for a year, then moved to the second floor of a private commercial building in Harvard Square. Our first full-time employees were Samuel Labate and William Lang; Sam stayed on, but Bill soon left to pursue his doctorate elsewhere.

In 1948, in *Slawinski v. J. H. Williams & Co.*, the State of New York declared that loss of hearing caused by loud noise was an occupational hazard and required compensation even though it might occur gradually over a period of years. This ruling produced a scare in the noisiest industries—drop-forge and metal-fabricating companies, for example—because they knew their employees had lost hearing acuity through exposure to intense noise on the job. Out of this came contracts to measure noise at one drop-forge factory and to make recommendations on reducing noise levels at others.

At this time, three brilliant students were working for their graduate degrees under Dick and me: Robert Newman in architecture, Jordan Baruch in electrical engineering, and Samuel Labate in acoustics (considered an interdisciplinary field at the time). Bolt and Beranek hired all three part-time before they earned their degrees.

Always well dressed, likeable, and full of anecdotes, Bob Newman had a master's degree in physics from the University of Texas and had worked a the Electro-Acoustics Laboratory at Harvard under my direction during the war. Afterward, he enrolled in the School of Architecture at MIT, where he was awarded his architect's degree in February 1950. After graduating, he was promptly hired as an instructor in architecture at MIT. Dick and I felt the firm would benefit if one of the partners were an architect, so we brought Bob into the partnership and renamed the firm "Bolt Beranek & Newman" (BBN). With Bolt in physics, Newman in architecture, and me in communication engineering, we ranged across a spectrum of knowledge and experience that would serve us well in succeeding years.

A small, wiry, balding man, Samuel Labate was a second-generation Italian-American, who had come to MIT after World War II to study mathematics. On meeting Bolt and me, Sam decided to do his master's thesis on acoustical materials. He was a clear thinker and well liked by his fellow workers.

The most brilliant of my students, Jordan Baruch came to MIT to become an electrical engineer, and had sailed through as a straight-A student. Jordan was of athletic build, with a head of thick black hair. He seemed to know everything, and was quick to offer anybody help on an eclectic array of subjects from health and gardening to automobiles and electronic equipment. He joined the firm full-time after earning his doctorate in 1951.

As we got to be better known, a number of architects called on us to advise them on the acoustical design of auditoriums, often in schools. One embarrassing situation arose with our design of a high school auditorium in Maine. We decided that the three of us would sit together and make a "best" design, which meant spending time well in excess of what an architect would pay. The architect was elated by our result and, when built, the auditorium earned plaudits for acoustical excellence. Within a year, we received a similar request from an architect in Texas, the space in question being almost identical in size and shape to that of the Maine project. We decided simply to send him the Maine design and instructed our secretary to copy the Maine report, taking care to change the location to Texas. A week later the Texas architect responded: "I am very pleased with your design. Attached to your mailing was the report that you copied from, which I am returning." No harm done, but we felt chagrined to have been caught in this act of self-plagiarism.

Two topics—how to reduce noise from heating, ventilating, and air-conditioning (HVAC) equipment and how to achieve speech privacy between offices—grabbed my attention, partly because so little had been done on them. There was nothing in the literature about acceptable noise levels in office spaces. As described in chapter 4, I surveyed office spaces at MIT, the General Radio Company, and a metal fabricating company in Troy, New York, questioning workers to learn how they felt about noise levels in their workplace and measuring these with special sound level meters and spectrum analyzers. The survey yielded promising results, and we applied for government funding to conduct a more extensive study.

The U.S. Air Force awarded us a contract that took me to Hill Air Force Base in Colorado where office noise levels varied over a wide range of levels and time periods. I published the results of this work in the *Journal of the Acoustical Society of America* in July 1956. With minor adjustments, the article has set the standard for acceptable noise levels in office spaces worldwide.

With the arrival of the commercial jet age in the 1950s, Bolt Beranek & Newman moved into new lines of expertise. The Pratt and Whitney Company of Hartford, Connecticut, asked us to quiet the cells used in testing jet engines. To my amazement, the company was testing these engines outdoors, mounted on stands, which not only produced an intolerable noise but blew dust into the neighboring residential area as well. Not surprisingly, it had been inundated by complaints. We helped plan test cells 10 feet square and 50 feet long, on each end of which were towers that opened upward; air for the engine was thus sucked in at one end and the exhaust discharged vertically through the other. We also installed noise-reducing baffles in the two towers. Our designs worked so well that we soon found ourselves in demand to consult on test cells for other engine manufacturers.

A Key Opportunity

Then came an exciting project that helped put Bolt Beranek & Newman on the map. I was in Washington, D.C., on Wednesday, January 18, 1950, testifying before a congressional committee about aviation noise at military bases when I was handed a note from my office to call the director of the NACA

(National Advisory Committee for Aeronautics; the forerunner of NASA) Lewis Flight Propulsion Laboratory in Cleveland as soon as possible.

He was frantic. At about midnight two days earlier, the laboratory had put into operation a new jet engine in a supersonic wind tunnel. The noise produced was so intense that switchboards in police and fire stations, radio stations, and public offices lit up nonstop with complaints from neighbors. The noise sounded like a series of thunderous explosions, even at distances as far as five or ten miles away. And the noise was the least of it—what really got to people was the vibration, which shook them and rattled windows. The city ordered an immediate halt to the operation, although permission was granted for the facility to operate for an hour or two the following Saturday afternoon for data collection. The long and short of it: the director appealed for BBN to send a noise measurement crew that Saturday, and pleaded, further, that we take on the design of whatever was necessary to mute the noise.

When I arrived at the Cleveland facility Friday morning to learn more, I was hurried to a room in which fifty or more people huddled around the largest conference table I had ever seen. The director opened the meeting in solemn tones, laying out a grave picture. The shutdown of the jet test facility, he said, was so serious that if the problem could not be solved, the future of the entire laboratory was threatened. I replied that BBN's noise-measuring crew would arrive the next morning and that we could handle the job, provided that the lab would agree to two conditions. The first was that I could only be in Cleveland Friday through Sunday each week, as I had to teach Monday through Thursday. This meant that the relevant engineering crew would need to work with me on weekends. My second condition was that the lab must build a model of the tunnel adequate for testing the noise reduction measures that we devised. The director went around the table seeking confirmation and, after some hasty discussion, turned to me and said: "We agree to the requirements you have laid out. Let's work out financial matters separately—for now, we'll expect you and your crew at the gate tomorrow morning at 9 AM sharp."

Briefly, the facility allowed for a jet engine to be mounted in a test section— 8 feet high by 6 feet wide and 40 feet long. An air stream created by huge fans was blown past the engine at supersonic speeds, just as though the engine was on an airplane flying supersonically. The speeding air, combined with engine exhaust fumes, was fed into a metal conical tube over 200 feet long, the large end of which was 26 feet in diameter. In other words, the small end of the conical tube had an area equal to that of the test section—48 square feet—and the circular large end measured 531 square feet. The conical tube caused the air to expand slowly before entering the atmosphere.

Two of our staff arrived with measuring equipment on Saturday morning. We loaded it into a van and prepared for the tests to start promptly at noon. We took measurements along a line out to a distance of 3,500 feet. The measurements indicated a need for about 30 decibels of quieting between 5 and 300 hertz (cycles per second) and about 20 decibels above 300 hertz if the noise were not to disturb neighbors. Such a level of noise reduction was enormous and had never been attempted before in a large facility. The design was further complicated by NACA engineers' requirement that the added muffling must not detectably impede the flow of outgoing air.

Guided by my specifications, the engineers constructed a 12:1 scale model of the tunnel in which provision was made for blowing air at high speed through the test section. The equivalent of the jet noise was introduced by a loudspeaker to the side of the test section. By a fortuitous coincidence, I had just completed a research project at MIT in which I had discovered a new method for high noise reduction in a duct, and I chose this method for the 12 to 800 hertz region. To attenuate frequencies down to 5 hertz, I selected Helmholz resonators which had to be installed in the sides of the 200-feet-long conical tube. At frequencies above 800 hertz, customary engine test cell noise-reducing treatments were chosen. All three components were first tested using the scale model. This was the equivalent, at full size, of a muffler about 220 feet long, 33 feet wide, and 46 feet high.

It was tough going. I spent near-sleepless nights planning the configuration and calculating specs. I discussed the design with Dick Bolt and Uno Ingaard, a physics-acoustics colleague at MIT. Ingaard doubted that the attenuation needed at the lowest frequencies could be obtained using resonators, and Bolt worried about the adequacy of my new treatment for the middle frequencies (12 to 800 hertz). Ingaar went so far as to suggest that the Helmholtz resonators would provide no noise reduction whatsoever. I listened to their arguments, but was not persuaded. I trusted my own instincts and the model, fortunately, allowed for comprehensive vetting prior to construction on a "real-life" scale.

The design and all testing with the model were completed within seven months of our first meeting, construction began, and the lab resumed operation within a year. As I expected, the muffled supersonic test facility was so quiet that neighbors were not even aware it was up and running. The structure, heralded as the "world's largest muffler," was featured with full-page photographs in *Life* magazine.

A Reputation in Ascendance

The United Nations project demanded a great deal of my time. And it involved not just one structure, but many—the General Assembly building; another building with three other auditoriums, one of which to be used by the Security Council; and, finally, office spaces. The principal architect was Wallace Harrison. The main part of the General Assembly is a large truncated cone standing upright. The delegates sit at tables on the ground floor of the cone facing the north "wall." On the sound side is a large two-level seating space for an audience. The design of the sound system for the hall was left up to me, and it proved almost undoable. Using the strongest language appropriate, I tried to convince Harrison that the loudspeakers should be hung directly above the main podium, encased in an acoustically transparent housing, perhaps looking like a world globe. He would not even consider this, and instead insisted that the loudspeakers be imbedded in the north wall of the cone. This meant that they would be about 25 feet behind the main podium and 25 feet above it, a sure prescription for howling feedback because the amplified sound had to be strong enough to reach the farthest seat in the audience, 165 feet away. With no choice in the manner, I had to try to make it all work somehow.

Fortunately, I had attended a lecture by a vice president of the Altec Company a few weeks before. Using actual equipment, the vice president had demonstrated that a certain kind of microphone could be placed directly in front of an Altec loudspeaker and intelligible speech projected without howling feedback. Blindly, with no time for exhaustive tests on full-sized equipment, I took a chance and chose the same microphone and the best loudspeakers made by Altec. I eliminated acoustical resonances in the spaces around the imbedded loudspeakers by packing them with sound-absorbing blankets.

Almost no one thought this system would work. The "behind the podium" location of the loudspeakers was so controversial that the UN director of telecommunications resigned and returned to Sweden, his home country, because he thought a disaster was in the making and he did not want to see it happen on his watch. But, happily, the system worked without any feedback and speakers' voices were perfectly intelligible in the most remote seats.

The success of this project spread Bolt Beranek & Newman's name far and wide, and our business boomed. In 1951, we moved to another Harvard Square building, at 16 Eliot Street; by 1955, we occupied the entire building. Jordan Baruch and Samuel Labate were admitted to the partnership in January 1952; from then on, BBN had five equal partners. On December 31, 1953, the company was incorporated, chiefly to protect the partners from liability that might arise in an important area of our business: control of jet aircraft noise. Bolt was named chairman of the board, I became president and CEO, Labate executive vice president, Newman vice president, and Baruch treasurer. At the time of incorporation, Bolt Beranek & Newman had 39 employees, 26 of them full time.

Whether we worked on government contracts or for independent architects, BBN was compensated on the basis of hours spent. Time sheets were mandatory, and proved crucial in gauging staff performance. At one point, I discovered with some alarm that only about half of each worker's 40-hour work week was chargeable to clients. To keep a closer watch on how employees were spending their time, I moved my main office from MIT to BBN and reduced my commitment at MIT to halftime.

In June 1954, Bolt Beranek & Newman accepted an assignment from the Convair Aircraft Company in San Diego to quiet Convair Model 340, a two-engine, propeller-driven, 44-seat passenger aircraft, already in production. The noise in the passenger compartment was so intense that airlines were not buying it. Because time was of the essence, we had to work on-site at the factory with Convair's engineers. To collaborate with me on the job I chose Edward Kerwin, who had just joined BBN after earning his doctorate in acoustics at MIT. We rented an apartment in La Jolla, where we lived from July 15 through August 30. Our wives came with us. We ate out every evening and found San Diego cuisine splendid. On five successive weekends, we visited Los Angeles and San Francisco twice and Tijuana, Mexico, once.

We had excellent cooperation from the Convair engineers. An office was assigned to us, and we had access to all necessary information. The airplane had two propellers driven by piston engines. I designed a muffler based, as in Cleveland, on my laboratory findings at MIT. Kerwin came up with the idea of rearranging the exhaust tubes to reduce muffler size. We also recommended lining the interior of the fuselage with an acoustical blanket, and installing double windows. The airplane turned out to be even quieter than expected. Convair was elated, and sold over 100 propeller-driven 340s before the jet age arrived.

Where to Go with BBN: Options and Challenges in the Digital Age

By 1957, our office at 16 Eliot Street was bursting at the seams. Some partners and employees were strongly opposed to further growth, others were eager to see the company expand. As chief executive officer, I made the decision in January 1957 to move to a substantially larger space—a one-story brick building at 50 Moulton Street. Soon afterward, I started to worry that, with the financial profits from their work all accruing to us, we could face a high turnover among our higher-level staff members. In 1958, I took several steps to ward off this possibility. We tied salaries by formula to gross fees and established an employee stock purchase plan. We designed a promotion structure for technical personnel parallel to the standard corporate ladder. Our new job titles were "Consultant," "Senior Consultant," "Principal Consultant," and "Chief Consultant," with the title "Consultant" replaced as needed by "Engineer," "Physicist," or "Psychologist." We set up a pension trust in 1954. In 1961, we created a life insurance plan and offered medical and hospital coverage. Employee response was positive and, from then on, we were in a good position to retain top people. I believe our company motto also helped motivate our staff: "Each new person hired should raise the average level of competence of the firm." In my experience, people of quality like to be surrounded by others who are as capable as they are, or more so, and who can inspire in them new heights of expertise and accomplishment.

Around 1955, I began to think deeply about Bolt Beranek & Newman's long-range prospects, guided in this by my wartime experience with the Electro-Acoustic and Psycho-Acoustic Laboratories at Harvard and the Systems Research Laboratory in Jamestown, Rhode Island. I visualized a potential growth area for BBN as man-machine systems that efficiently amplify human labor. Two such systems—optimization of blind landings at airports and improvement of racing boat performance, as in the America's Cup—kept me awake at night sometimes, mulling over the remarkable possibilities.

I reflected on the people working in related areas, and J. C. R. Licklider—the colleague in electrical engineering at MIT who had helped us out in the Hush-A-Phone litigation—came to mind. Not only was he a first-rate psychologist with physics training, but he had also learned much about the uses of computers through his exposure to the air defense system SAGE and other pioneering work of the computer masters at MIT's Lincoln Laboratory. My appointment book shows that I courted Licklider with numerous lunches in the spring of 1956. At a meeting in Los Angeles that summer, I asked him outright to join BBN and to assume leadership of a new direction with us. Because a professorship at MIT was not something Licklider (or anyone else, for that matter) would give up lightly, I knew I had to make a compelling case. So I played up not only the career and research opportunities at BBN but also Licklider's discomfort with the MIT administration on a key matter. I was aware, I told him, that MIT had not favored his desire to form a psychology department, and I promised that BBN would more nearly satisfy this goal. I offered him a large stock option at current book value, $1.50 a share, and the title "Vice President in Charge of Man-Machine and Information Systems." I emphasized that Bolt Beranek & Newman was well known for hiring technical people of the highest caliber, and that he would find himself surrounded by leaders in their respective fields, in other words, that he would find at BBN as stimulating an environment as he had grown accustomed to at MIT.

"Lick," as he insisted we call him, came aboard in the spring of 1957. A thin-boned, almost fragile man, with thinning brown hair and enthusiastic blue eyes, Lick walked with a gentle step, often with Coke in hand, and always found time to listen and share new ideas. He was relaxed and self-deprecating and would end many a sentence with a slight chuckle, even when he was dead serious. He mingled readily with the talented staffers already at BBN; he and I worked together especially well, even when we occasionally disagreed.

Lick had been on staff just a few months when he asked us to buy a digital computer for his group. I pointed out that we already had a punch card computer in the financial department and several analog computers in the artificial intelligence group. He was unimpressed. "I want a state-of-the-art digital machine," he said, "the LGP-30, produced by the Royal-McBee Company, a subsidiary of Royal typewriter." I asked him what it would cost. "Around $30,000," he replied rather matter-of-factly, adding that this price tag was a discount he had already negotiated. More than a little taken aback, I exclaimed that BBN had never spent anything close to that amount on a single piece of equipment. And when I recovered enough to ask him what he proposed to do with it, he responded: "I don't know, but if BBN is going to be an important company in the future, it must be in computers."

Having great faith in Lick, I finally came around to the notion that spending $30,000 for something with no immediate use was well worth the risk. I carried his request to my partners, and, with their concurrence, Lick brought BBN into the digital era.

He sat at the LGP-30 many hours each day, monopolizing the machine as he taught himself digital programming; over the next several months, he hired key personnel interested in artificial intelligence and man-machine systems. By the time we printed our 1958 client brochure, Lick's department consisted of two divisions, one to identify how best to establish and control information flow, whether between humans or between humans and machines, and a second division to determine engineering criteria for an optimum man-machine system, be it a factory, a vehicle, or a computer.

Within a year of the arrival of our LGP-30, Ken Olsen, president of the fledgling Digital Equipment Corporation (DEC), stopped by Bolt Beranek & Newman, ostensibly to see our new computer. After satisfying himself that Lick really understood digital computation, he asked if we would consider taking on a new project. DEC had just completed a prototype of its first computer, the PDP-1, and needed someone to try it out, for a month. We agreed to do just that.

The prototype PDP-1 was a monster compared to our LGP-30. It would fit nowhere in our building except the visitors' lobby, where we did our best to obscure it behind elegant Japanese screens. Lick and Ed Fredkin, an eccentric young genius who came to BBN to work with the LGP-30, along with several other staffers, put the PDP-1 through its paces, after which Lick came up with a list for Olsen of suggested improvements to make it more user-friendly and more effective.

The new computer won us all over. In 1960, I arranged for DEC to provide us with their first production PDP-1 for $150,000; Lick and I immediately left for Washington, D.C., to seek out research projects that would put the machine to good use. Our visits to the Department of Education, National Institutes of Health, National Science Foundation, and Department of Defense resulted in several lucrative software contracts—clear proof that computers were indeed the wave of the future.

As BBN president, more and more of my time was consumed by company activities. I had reduced my load at MIT to half-time in 1953 and finally resigned in 1958. Bolt remained at MIT as a full-time professor, continuing to devote his "day a week" to BBN. Newman committed himself half-time to BBN, while retaining part-time appointments at MIT and Harvard.

In the spring of 1959, I took Ira Dyer, head of BBN's applied physics department and an MIT Ph.D., to Washington to see if we could acquire a contract with the newly established National Aeronautics and Space Administration (NASA). When we arrived, we were sent over to see a scientist, whom we found in an odd-shaped room dominated by a huge blackboard completely scribbled over with mathematical symbols. He was having trouble, he told us, solving an equation vitally important to his work. Dyer studied the chalk scribbles for a moment or two, then piped up: "I see a solution—let me show you!" His answer was correct. The NASA scientist was amazed and warmed up to Dyer right away. From that point on, we never had any trouble acquiring NASA contracts, in particular, contracts for the control of sound and vibration in space vehicles.

Lick and I worked hard to secure government support for research on speech compression, on criteria for prediction of speech intelligibility in noise, and last but not least, on the reaction of communities around airports to aircraft noise. We were able to hire Karl Kryter, whom I had known and Lick had worked with at the Harvard Psycho-Acoustic Laboratory during World War II. An expert in speech bandwidth compression and the effects of noise on sleep, Kryter soon became part of our team for the Port of New York Authority's project on jet noise reduction at the major metropolitan airports (described in chapter 6).

Walter Juda came to us from Ionics, Inc., a company marketing machines for converting brackish into potable water. He was convinced that fuel cells, which were highly efficient and emitted only water vapor as a by-product would be the next major source of power. When we hired Juda to do research in that field, General George Dorio, Boston's venture capital leader at the time, who had financed Ionics and considered Juda to be the key person there, was furious. He called me on the phone and almost ordered me over to his office. I went with Jordan Baruch, having learned that, where legal action might ensue, a witness was essential. Sure enough, Dorio had a lawyer with him when we arrived. The lawyer warned us that his client was prepared to sue BBN for theft of Ionics ideas, through the hiring of Juda, but would refrain from doing so if I persuaded Juda to return to Ionics. My response was swift and blunt: "Go ahead and sue. We're not going to use any of Ionics ideas—Juda will work for us in another area." With that, Jordan and I got up and walked out. No suit was ever filed.

Bolt Beranek & Newman grew without the help of outside financing, except for a line of credit at the First National Bank of Boston. With a debt of $325,000 in 1961, however, it was apparent that we needed further cash for expansion. So we went public. Working with our auditors and lawyers, as treasurer and CEO, Jordan and I planned BBN's initial public offering; made on June 27, 1961, it raised nearly $1 million. In selecting an underwriter, we interviewed several investment firms. Paine-Webber thought our offering price should be $4.50 per share; Smith-Barney thought $8.50. We chose Hemphill Noyes & Co., which took us public at $12 per share. The price rose to $18 on opening day, and remained above $12 well beyond the next year.

With the arrival of our PDP-1, Lick hired two MIT professors—John McCarthy and Marvin Minsky—as consultants on computer programming, nominally to help with our work in artificial intelligence. Seeing two strange men in one of our conference rooms in the summer of 1960, I stopped by and asked who they were. "Well, who are *you?*" McCarthy countered. When the three of us got acquainted, I realized we had a pair of geniuses in our midst.

McCarthy had developed the concept of computer time-sharing at MIT. Despite his urging, however, the computer people there were slow to implement it. He found Lick and especially Ed Fredkin of BBN to be more engaged and more responsive. Fredkin insisted that time-sharing could be done on our PDP-1. But McCarthy kept insisting that an interrupt system and a swapper were needed to suspend computations in progress and to perform exchanges among computational streams, respectively. Fredkin said we could modify the PDP-1 to manage both.

Led by Sheldon Boilen, our team divided the PDP-1's operation into four parts, assigning each to a separate user. In the fall of 1962, we conducted a public demonstration of time-sharing, with one operator in Washington, D.C., and two in Cambridge. To augment the small memory of the PDP-1, we had acquired the first FASTRAND rotating drum, made by UNIVAC, with 45 megabytes storage capacity and an access time of about 0.1 second. The success of this demonstration solidified our position as computer experts.

With financing from the National Institutes of Health, we installed a primitive time-shared information system in the winter of 1962 in the Massachusetts General Hospital. With Jordan Baruch as its parent, the system allowed nurses and doctors to create and access patient records at a number of hospital stations, all connected to our central computer. This was the first use of computers for managing patients in hospitals.

The time-sharing idea worked out so well in this instance that we set up a separate time-sharing business, TELCOMP, in 1965, using a later model of DEC computers. By dialing our computer's number and then using tele-typewriters we provided, subscribers who were usually research engineers in Greater Boston companies or universities, would gain access to its input on a time-sharing basis. TELCOMP grew in New England and soon expanded into New York; by the end of 1965, $1 million had been invested in the enterprise. An important milestone was reached in 1970, when we signed a comprehensive contract with Harvard University. The system became Harvard's exclusive source of time-sharing services, encompassing university-wide functions in civil and structural engineering, electrical engineering, financial planning and analysis, and business management.

But, just as TELCOMP appeared poised to go national, disaster struck. General Electric had interested Dartmouth College in adapting for time-sharing one of a line of digital computers it could not sell in competition with DEC. When the plan worked out well, the company then set up its own time-sharing business, using the same line of computers at greatly reduced access rates. GE had already written off the computers as valueless, hence there was no depreciation cost in offering the service. We had little choice but to bow out of that business.

Then, in the summer of 1962, Lick was lured away by Jack Ruina, director of the Advanced Research Projects Agency, to head ARPA's Information-Processing Techniques Office in Washington that fall. I tried my best to keep Lick at Bolt Beranek & Newman. "You've become famous by building a computer sciences department here," I argued, "and a civil service job just doesn't seem right for you somehow." But he had thought the decision over carefully. The government promised him substantial funds that he could distribute to universities to set up their own time-sharing computer centers. Moreover, thanks to the profit on his BBN stock following our public offering, he could afford to take a chance on his future. So we lost him. Lick completed his 21st Century Library Project report, begun at BBN, in the form of a book, *Libraries of the Future,* with chapter contributions by five other BBN experts. The book achieved national recognition, and was influential in pioneering the use of computers in libraries.

A Foreign Subsidiary

In the early fall of 1958, I visited the Technical Institute of the University of Berlin, headed at the time by an acoustical engineer—Professor Lothar

Cremer—who had worked for a year at Bolt Beranek & Newman. He introduced me to his graduate students and I took a particular liking to Manfred Heckl, then working toward his doctorate. Speaking privately with him over lunch, I learned that Heckl hoped to finish his work that academic year, but that Cremer wanted him to spend another year. Impressed as I was with Heckl, I went on about BBN at some length and suggested that he join us for a while. He jumped at the offer.

When next I saw Cremer, I told him of our conversation and said it would be good for acoustics in both countries if Heckl could work with us in Cambridge for a few years. Although he did not answer right away, in due course, Cremer informed us that Heckl would receive his doctorate at the end of the 1958–59 academic year after all. Heckl brought his family to Cambridge and remained with us for three years.

In Heckl's second year at BBN, Labate and I talked with him about forming a similar consulting firm in Munich. He told us about his close friend Helmut Müller, who was running an acoustical consulting business in Munich in partnership with Cremer. Heckl got Müller to come to Cambridge to talk with us. In the end, Müller, Heckl, and Cremer, along with Müller's right-hand man Ludwig Schreiber, joined with us in setting up a German branch of BBN.

With most of the financing provided by Bolt Beranek & Newman, on June 15, 1962, Müller-BBN GmbH was born. I was named chairman and retained that position for many years. Every spring, I traveled to Munich to oversee the annual meeting. Although the company was always profitable, in 1972, BBN decided that maintaining outposts in other countries was not in its best interests. The principals (by then I had retired from active management at BBN) offered their Müller-BBN stock for sale. I bought a portion and the balance was sold to the employees of a renamed company "Müller-BBM."

Back to School

Bolt Beranek & Newman was growing to the point where I felt I needed to broaden and deepen my management skills. I had heard favorable reports about the Advanced Management Program at Harvard's Business School, and decided to enroll for the spring of 1965 (the program took place over a 13-week period each spring and fall). Much to my surprise, attendees had to move into a campus dormitory. Having thought I'd left the communal lifestyle of college far behind me, I arrived at the Soldier's Field campus with luggage in tow one chilly day in February 1965.

Already milling about were some of the other 160 "students," if we could be called that at our relatively advanced time in life, 38 of whom came from foreign countries. Our assigned dormitory was Hamilton Hall, where the class was divided into 20 "Cans"—groups of eight living together in several rooms with four two-level bunks, eight desks, and a shared bathroom (hence the group name). My group—Can 8—consisted of Clarence Michel from a London building company, Chet Goss from CIT, Bill Dunkle from United Airlines, Paul Joy from Carborundum, Jim Tracy from Standard Oil, Fred Shiavi (engineer) from the U.S. Navy, and Ike Joaquin from IIT Philippines. Goss brought an electric hot plate and often cooked up a pot of chili con carne for the group. Dunkle seemed more interested than the rest of us in socializing. Joaquin stayed out most nights, slept during the day, and hardly ever studied.

Classes convened in a large, fan-shaped auditorium, with rows of desks perched on terraces; small conference rooms were available for group discussions. A principal feature of the program was the opportunity to get to know practicing professionals from a range of backgrounds, cultures, and business settings. The program's purpose, Dean George Baker explained, was "to prepare men to assume responsibility for top leadership in organizations and to exercise these responsibilities in ways that serve well their organizations and society at large."

The "case method"—a hallmark of the Harvard Business School curriculum—was used in each class. A case was usually the result of an in-depth study of a company carried out by one or more professors, who would analyze problems that had been successfully or unsuccessfully addressed by managers on-site. The question for us students, invariably, was: "How would *you* solve the problem?" We had a half dozen professors for marketing, business policy, financial policy, labor relations, and statistical decision making, all throwing cases at us left and right. But to my surprise, I found myself coming up with answers as fast as anyone else in the class—and sometimes faster.

One baffling case comes to mind as an example. In a medium-sized company, confidential discussions at board and management meetings were being leaked to other employees a few hours after the conclusion of each meeting. Even though attendees were asked to maintain secrecy, the leaks continued unabated, and it was a long time before anyone could figure out why. No one in our class had a clue either, but I came up with what seemed to me the obvious (and, as it turned out, correct) answer. The telephone operator (still a common functionary in those days) had listened in on calls made among staff members freshly returned from the meetings and then passed the information along to all and sundry. I found, in fact, that my experience and exposure were broader than most in the class. I was the only one who had taken a company public, for example, and I was pretty much alone in my knowledge of digital computers.

I served as ad hoc secretary for the group meetings held once a week to discuss our experiences and to make suggestions to the faculty. I learned, after arrival, of a long-standing custom—that each enrollee either give a present or do something for the other 159. John Randall from Corning Glass presented each of us with a piece of Stuben Glass. Bill Dunkle took us soaring over Niagara Falls in a Boeing aircraft. Art Knight, of the Mystic Steamship Corporation, whisked us on a cruise around Boston Harbor. With little money to spend, I could hardly match any of these—but I did hire two buses to take anyone interested on a tour of BBN's offices on Moulton Street, just a few minutes away and about a third came.

The high point of the course was an assignment given to all the Cans. Each Can had to compose a consulting report and make an oral presentation before the entire class on how to improve the operation of the General Foods Corporation's International Division (GFID), headquartered in New York. For background, we were furnished with a report prepared by three staff members of the Harvard Business School. Our group suspected that most of the other Cans would focus on marketing and acquisition policy. Wanting to be different, we focused on management training in GFID's foreign operations (the division that ran manufacturing and marketing operations in 13 countries outside the United States and Canada) and on intra-firm communication.

About half our report was devoted to an electronic data-processing system for management and control. We pointed out that, at GFID, cost accounting was almost unknown and there were excessive delays between the generation and collection of basic financial and business data from overseas operations. We recommended that the division create a time-shared, real-time electronic data-processing system for all its foreign operations, with information transmitted at regular (nighttime) intervals by satellite to the home office. We predicted that, if adopted, the division would see a substantial increase in profits owing to a speedup in information gathering and processing, improved decision making, and reduced staff.

Can 8 chose me as report writer and presenter. When I finished speaking, my group cheered. We thought we were a shoo-in for top honors, but marketing won out in the end and our presentation only got an honorable mention. Good enough, we conceded, but still disappointing for a group of highly competitive, gung ho fellows wanting to bask in the glory of "number 1."

There was a thrill to the politics of the program. At the end, a nominating committee produced a ballot for class officers, with two contestants for each of three offices: chairman, secretary, and treasurer. I was nominated for secretary. The losing nominees would be designated deputy chairman, assistant secretary, and assistant treasurer. My rival was Art Knight. Although I bought drinks on several occasions, printed napkins emblazoned with the words "Vote for Beranek," and made the rounds of various Cans to drum up votes, Art's cruise of the harbor won the day. Plus, his personality was truly engaging—more so than mine, I suspect. All the same, it was pleasant to realize that out of that forbiddingly talented array of 160 entrepreneurs, I was one of just six elected to represent the entire class permanently. The Advanced Management Program was helpful and stayed with me over the years. I returned to BBN full of ideas and more secure in the skills of analysis, communication, and careful presentation.

Origins of the Internet

In the 1960s, two of Bolt Beranek & Newman's software projects needed special expertise: the hospital time-share project and a computer system for company-wide use in a large firm in suburban Boston. Dick Bolt convinced Frank Heart, who was working at MIT's Lincoln Laboratory, to come aboard as head of our information sciences and computer systems division. Frank quickly became one of BBN's key employees. Starting from the time of his arrival in December 1966, he helped us earn the sobriquet "Cambridge's Third University." Frank was the only software expert I ever met who could both accurately estimate the length of time it would take to complete a proposed project and end up within budget.

Early in 1968, the Advanced Research Projects Agency of the Department of Defense (ARPA) issued a request for proposals to build a computer network that would connect together large computers at leading universities and laboratories. By this time, Lick had left ARPA and gone over to IBM, where he stayed for a short time before completing his varied career as a professor at MIT. ARPA hypothesized that a packet-switched network would also allow small research institutions or universities to access large-scale computers at major research centers and thus relieve ARPA of having to supply each laboratory it supported with a multimillion-dollar machine. We at BBN decided to enter the competition and we chose Frank Heart to manage our proposal. He put together the Interface Message Processor (IMP) Group, which planned the equipment and the software and prepared a proposal submitted in September. Thirteen companies competed; in the end, Bolt Beranek & Newman and Raytheon were the finalists. The decision came down at Christmastime: BBN had won the million-dollar contract. Senator Ted Kennedy sent a telegram on December 23, congratulating us "on winning the contract for the interfaith [sic] message processor."

The first interface message processor invented by BBN was completed and shipped to UCLA in September 1969. The second went to the Stanford Research Institute a month later, and the two machines were hooked together by a 350-mile leased 50-kilobit telephone line. On October 3, the ARPANET—predecessor to the Internet—carried the first message between two packet-switched computers in different cities: "LO," pronounced "ello." The ARPANET grew in size and was managed by BBN for two decades with government support. When it reached the size of 562 nodes on January 3, 1983, the government divided the system into two networks. With the TCP/IP protocol, invented by others and chosen instead of ours because it was superior, as the joining software, the Internet was born that day.

Experiments in Management

Finding that much time was being spent by individual researchers or consultants on financing problems and that tighter controls needed to be set on chargeable time, billing of clients, and communication with the financial office, I came up with a novel management concept sometime around 1957. To meet these growing and in some cases conflicting challenges, I set up a financial arm in parallel to the research arm. Under this scheme, each technical department was assigned a financial officer. This individual, whom I called a "facilitator," reported to two bosses: the head of the department to which he was assigned and the company's chief financial officer. If a researcher wanted to buy a piece of equipment or to set up a new research facility, he would sit down with the facilitator and outline his needs. The facilitator would work out specifications with him and, after obtaining approvals from management, attend to the purchasing details and arrange for placement of new equipment and any necessary modification of building space. Collecting a weekly time sheet from each employee in the department to which he was assigned, the facilitator would also monitor sick time and vacation time. His further duties were to track the progress of each project in relation to the terms of its contract, to keep abreast of deadlines and of potential late penalties, and to bill clients based on time sheets and contract terms.

In my view, this arrangement benefited us in two ways: it freed our technical people from red tape, giving them more time to tend to research and, on the financial side, it established a new level of accountability in contract provisions, deadlines, and billings.

My own management style turned out to be a curious mix, perhaps reflecting my mixed technical, administrative, and teaching background. At the start of my career at Bolt Beranek & Newman, I was senior in age and experience to all employees except Dick Bolt. As a result of my own research and that of my graduate students at MIT, I came in with new knowledge, generally at or near the forefront of some rapidly emerging technologies. I took the lead in a number of key projects and acted as a close partner with our consulting staff. I appointed Sam Labate to be responsible for day-to-day management and I kept in constant touch with him.

Overall, my style involved working arm in arm, or face-to-face, with staff whenever possible, treating them as colleagues and building their pride in BBN as a professional organization of the highest caliber. I held weekly meetings with senior staffers to learn what they thought needed doing to improve our operations and output. I encouraged everyone to join the appropriate technical societies and to write papers for publication.

My tenure as president of Bolt Beranek & Newman ended June 30, 1969, at which time I was named chief scientist and general manager of consulting and of research and development. Sam Labate succeeded me as president and CEO. This transition came about somewhat unexpectedly.

In December 1962, I had joined a group of thirty men and women interested in obtaining a license for the operation of Boston's Channel 5, a large network-affiliated television station. In 1963, on the application that our group made to the FCC, I had agreed to serve as president, with the expectation that the executive vice president, Nathan David, would take over as president should Boston Broadcasters, Inc. (BBI), get the license. But David found himself embroiled in a questionable case of stock dealing, and had to resign. So I was stuck with a new career. Following extensive media publicity about the TV station, identifying me as BBI's president, I was pressured by the BBN board to resign as president and CEO. Sam had tried to get Boston Broadcasters to name another president so that I could remain at BBN, but BBI's board felt that this might suggest their organization was falling apart. This was a little over two years before a favorable U.S. Supreme Court ruling finally permitted us to go on the air. As we worked toward getting the station established in mid-1971, I resigned from BBN altogether, convinced that even a part-time consulting role there would no longer be feasible. After nearly a year of hiring and construction, Boston Broadcasters went on the air in March 1972 as WCVB–Channel 5–Boston, with ABC as its affiliated network (see chapter 8 for an account of this episode).

My departure was, I believe, in the best interests of the firm I helped found. I doubt I could have managed the digital network business as well as succeeding presidents have done. Bolt Beranek & Newman, in effect, kept plunging deeper and deeper into computers, the wave of the future. I had kept up with the emerging technology to a point, but it as neither my area of expertise nor my primary interest, and I didn't care to spend the rest of my career in the hard sell of building and marketing nationwide computer systems and networks.

6 Muffling the Jet Age

I was sitting in my office at MIT on a sunny day, November 7, 1956, when the telephone rang. "I'm Austin Tobin," the caller introduced himself, "director of the Port of New York Authority. We operate the airports around New York City. The PNYA is facing a potentially serious noise situation that we must deal with reasonably fast. We'd like to talk with you about it."

Once assured that I was not working for any airplane manufacturer on quieting or measuring jet engine noise or in any other way—"We want no conflict of interest," Tobin explained—he asked me whether I could meet with him in New York at his office at 4:00 PM the next day.

Taken aback, I replied: "I teach classes three days a week, and tomorrow's is from one to two. I've vowed not to be an absentee professor."

"Let me look in my flight guide," he said. "I see there's a flight leaving Boston at 2:30. It takes an hour to get to New York, and the weather will be good tomorrow. We'll have a helicopter waiting when your plane pulls up to the La Guardia terminal. Our building at 14th Street and 8th Avenue is the only one in New York on which a helicopter can land. My office is on the floor below the landing pad. We'll expect you at 4 o'clock."

My secretary arranged for the ticket. The next day I dismissed my class fifteen minutes early and raced out to MIT's main entrance, where my secretary was waiting with the car. There wasn't much traffic in those days, and at airports there were no ticket lines and no long walks to gates, so I managed to board the plane with five minutes to spare.

As we landed, I could see a helicopter with its rotor turning next to the American Airlines terminal. Someone on the tarmac called out my name. The helicopter was small, big enough for just three passengers including the pilot; the "bubble" enclosure was transparent, extending down to floor level. I've never had a more spectacular ride—low over the length of sunny Central Park and down Broadway. We landed on top of the Port Authority building, and I was directed down an outer flight of stairs. I arrived in Tobin's waiting room ten minutes early.

Joining Austin Tobin and me at the meeting were Matt Lukens, the assistant director of the Port Authority, John Wiley, the director of aviation, and a lawyer. Tobin was a stocky man of average height, about sixty years old, a little overweight, with thinning hair and a commanding presence; Lukens was younger, taller, more handsome, and sported a broad smile. Wiley, a cheerful man of fifty-five, trained at MIT, whom I'll always remember for his unfailing response to any difficulty—"no sweat"—would become my most frequent contact at the Authority.

"Pan American Airways," Tobin began, "has asked for permission to begin jet aircraft operations from Idlewild Airport in November 1958. They will be flying a new jet-propelled passenger airplane, called the 'Boeing 707.' We must know how noisy it is. We already have a lawsuit in connection with our Newark airport, brought on by residents complaining about takeoff and landing noise of propeller-driven airplanes. The noise surrounding Idlewild must not become worse than what these airplanes already cause."

"Several years ago," Wiley elaborated, "we told the airlines that a jet plane must make no more noise than a large propeller airplane on takeoffs and landings. Boeing claims they have met this requirement and their evidence is that the conventional sound level meter shows the same number of decibels of noise from the jet plane during flyover as from a Super-Constellation propeller airplane. But we're worried because we are getting reports that the 707 sounds much noisier than a Super-Constellation. In fact, we have already received threats from neighborhoods around Idlewild that if the noise increases, mothers with babies will march onto the runways in protest." He looked directly at me. We want you to establish a thorough program to investigate this problem so that we know what we're up against and what we should do about it."

They'd been impressed by what they'd heard and read about us, particularly by *Life* magazine's coverage of our successful design of the world's largest muffler—200 feet long and 30 feet square for NACA. I assured them we could handle the problem. The first step, I said, would be for the Authority to arrange for Boeing to make flyovers of the aircraft intended for delivery to Pan American and to let us make measurements of the noise, with Tobin and Wiley present. All four men agreed.

The First Boeing Tests

A few weeks later, we traveled together to Boeing headquarters in Seattle. Boeing had refitted the 707, formerly a military jet tanker plane, to serve as a passenger aircraft. But it had no mufflers. My measuring equipment consisted of the latest in microphones and an excellent portable magnetic tape recorder.

Boeing flew the prototype overhead at a height of 1,000 feet roughly the altitude of planes taking off over the houses closest to Idlewild Airport. We were stunned—the noise was terrible. Tobin, Wiley, and I agreed that it was completely unacceptable. "We can tell you right now," Tobin informed the Boeing people, "we will not approve the use of our airport by a plane this noisy." They bristled at this, pointing out that the noise made by the 707 and by the largest propeller airplanes measured the same on a sound level meter. "You have already said that this is the test for acceptability at Idlewild."

Boeing had a point here, but a superficial one. Although the sound meter measurements looked the same, there was a marked difference in "noisiness" created by the two types of aircraft. A propeller plane has its loudest noise in the low-frequency (low-tone) range of 50 to 2,000 hertz, whereas the jet plane has its loudest noise in the high-frequency (high-tone) range of 500 to 2,000 hertz. Human hearing is many times more sensitive to high- than to low-frequency noise, but the sound level meter treats high and low frequencies equally.

Back in New York, after explaining that Bolt Beranek & Newman had just employed one of the best experimental psychologists in the country, Karl Kryter, and that we had one of the finest noise-measuring crews with the latest equipment, headed by Laymon Miller, I made a proposal. "I will take charge of the project. Using test listeners, Kryter will compare the jet noise with propeller noise and determine the extent to which the jet noise would have to be reduced to meet the requirement of 'no more noisy than large propeller airplanes.' Miller will measure the propeller-aircraft noise off the ends of runways at Idlewild Airport and use those data as a basis for comparison. If you authorize BBN to go ahead, the three of us will be responsible for the whole project and we are prepared to spend almost full time on it until completion."

We already knew, from the Newark airport lawsuit, the limit of neighbors' tolerance to daily propeller-aircraft operations. Our investigation started with measurements of the takeoff noise of propeller-driven aircraft in residential areas around Idlewild, particularly at distances between 2 and 3 miles from the start of takeoff roll. We also used cameras with high-quality lenses pointed upward to determine the height of each flight. From the control tower, the Port Authority was able to give us the type and ownership of each airplane, what its gross weight was on takeoff, who was piloting it, and the exact time its takeoff roll began.

Over the next six months, we learned that noise in a community off the end of a runway varied according to the make of the airplane, who was flying it, its load, and the weather. On a cool day or with partial load, planes flew higher and were less noisy. One airline's pilots always flew lower on takeoff than other airlines' pilots, making proportionately more noise. We took our measurements at all times of day, in order to determine in a comprehensive, rigorous way precisely what noise levels neighbors were experiencing.

The next step was for the Port Authority to arrange with Boeing to conduct fully loaded (with lead bars to simulate a full passenger load) 707 jet aircraft takeoffs using their Seattle airport runway. These tests were scheduled for just after New Year's Day 1957. Boeing tried to put us in a good mood by hosting a lavish dinner for us the night before. We were told how much the new airplane meant to the company and how it was certain to open up a new world of commercial travel. The journey by air from California to New York, they boasted, would be cut from eight hours down to six. But these lobbying tactics made little impression on the Port Authority people, who remained skeptical and content to await analysis of the test results.

Noise level measurements were taken and magnetic tape recordings made by Boeing's engineers and by BBN personnel. To avoid potential disputes about whose set of data was more accurate, I made sure the two teams compared instrument calibrations ahead of time. The tapes were brought back to BBN's laboratory so that Kryter could set up the psychoacoustic experiments to assess the relative "noisiness" of the Super-Constellation propeller airplane and the Boeing 707 jet. In a first set of tests, BBN engineers and secretaries, pressed into service as subjects, listened first to Super-Constellation, then to Boeing 707 noise at a typical level, as played back over loudspeakers. They were instructed: "Adjust the level of the jet noise, using a knob, until it sounds equally noisy [so that it] would be no more and no less disturbing to you in your home than the propeller aircraft noise." In a second set of tests, listener subjects from outside BBN were given somewhat modified instructions to eliminate any ambiguity. Later, a third set of tests was run using two groups of 50 people each at a large research laboratory on a U.S. Air Force Field. The thoroughness of our data thus seemed beyond reproach.

All four groups of listener subjects judged that, when measured with the standard sound level meter, the noise levels of the 707 jet, fully loaded, had to be reduced by more than 15 decibels if they were to be no more disturbing than the noise levels of a Super-Constellation propeller-driven aircraft, fully loaded. Such a reduction was enormous—tantamount to reducing the noise made by 30 jet engines to that made by a single engine. The Port Authority carried this information to Boeing and the company was compelled to design and install heavy mufflers for the exhausts of the four engines on the 707.

A New Assessment Scale: Perceived Noise Level in Decibels

Our man Kryter then developed a method for calculating the subjective "noisiness" of aircraft, whether jet or propeller-driven, as actually perceived by humans. Taking into account that the human ear is less sensitive to low- than to high-frequency sounds, Kryter's method arrived at a new measurement, called "perceived noise level in decibels," or PNdB. Equal PNdB readings meant equal "noisiness" as judged subjectively. Taking the extensive propeller-aircraft noise data recorded by Laymon Miller's group off the end of the runways at Idlewild, where the nearest residence was in the city of Howard Beach about 2.5 miles away, Kryter converted them from objective to perceived noise levels in decibels. About a fourth of the takeoffs measured by the group exceeded 112 PNdB. We reported these results to Austin Tobin and John Wiley.

French and British Aircraft

Meanwhile, quite coincidentally, a related and entirely unexpected request had come in from overseas. The French notified the New York Port Authority that they wanted to fly between Paris and Idlewild using their new Caravelle jet airplane. Tobin bluntly told them they "must meet the requirement that the plane not make more noise, subjectively, over communities around Idlewild than present-day, large, propeller-driven aircraft." The French promptly invited the Port Authority to send a crew to Orly Airport, outside Paris, to measure the noise.

Laymon Miller and I carried out these tests March 19–20, 1957, with the help of French engineers. We set up measuring stations at different distances from the start of the takeoff roll. The important new concept, special to the Caravelle tests, was that, after lifting off the runway, the airplane would climb very steeply at full engine power to an altitude of 1,200 feet, then cut back for a more leisurely, hence less noisy ascent. We converted the recorded noise data to PNdB. With the steep climb and power-cutback procedure, the "perceived noisiness" of the higher frequency noise of the Caravelle and the lower frequency noise of the prop planes at 2.5 miles were nearly identical. As BBN reported to the Port Authority: "It is our conclusion that the total Caravelle noise output is approximately comparable as far as listeners are concerned to that of large present-day four-engine propeller-driven airliners when flown as demonstrated." The Caravelle was, therefore, granted permission to use the airport regularly, provided it adhered to that takeoff procedure.

In November 1957, Laymon Miller and Bob Hoover measured the noise characteristics of the British Comet 4 in England. With the same takeoff procedure, this aircraft also met the Port Authority standard of being no more noisy than large propeller airplanes at the distance of the nearest Idlewild residents.

Boeing Again, and Now Pan American

After Boeing had equipped the prototype 707 with their best design of mufflers, we were invited back to Seattle. The tests of the fully loaded aircraft (again with lead bars as stand-ins for passengers) were conducted between April 21 and 24, 1958. The pilots used normal takeoff procedures. The noise levels in PNdB were still much larger, by about 6 decibels, than those for large propeller-driven airplanes. This difference is equivalent to shutting off three of the four jet engines on the 707. The plane did not come close to meeting the Port Authority noise requirements, and Boeing was so informed. The Authority asked us to make a demonstration tape, showing the difference in "noisiness," as heard indoors with open windows by a 707 jet with mufflers and by a Super-Constellation, both following standard takeoff profiles. This tape made it clear to listeners that the difference was enormous.

A few months later, Austin Tobin, John Wiley, Matt Lukens, and another Port Authority official took me and Kryter along on a visit to heads of airports in Germany, the Netherlands (Amsterdam), and France (Paris). We carried along loudspeakers and the demonstration tape. After hearing the demonstration, our hosts appeared alarmed about the impending jet invasion;

the immensity of the problem, it seemed, was registering with them for the first time.

A tense political moment arose, incidentally, in Germany. To demonstrate the effect of noise on voice communication inside a home, Kryter had selected speech from a phonograph record. Unfortunately, he chose a recording of the controversial Nuremburg tribunals of the late 1940s. The Germans, still wincing from the humiliation of their wartime defeat and its aftermath, were warned of this gaffe a few minutes in advance; a walkout was averted, but not before several participants had had an opportunity to vent their feelings about the proceedings at Nuremburg.

Boeing and Pan American were outraged when they learned about our demonstrations across the Atlantic. They had hoped to get other airports, both nationally and overseas, to accept higher noise levels and thus exert pressure on the Port Authority to follow suit.

In desperation, Boeing convened a meeting in Seattle on July 15, 1958, to which all airport heads, airline executives, Douglas Aircraft Company, and government aviation regulators were invited. Boeing's intent was to show that BBN did not have a solid basis on which to recommend to the Port Authority the use of perceived noise levels in decibels (PNdB) for evaluating aircraft noise. I stood at a blackboard in an auditorium before about 150 people almost all day, with only a break for lunch. I described how Kryter had arrived at the 15-decibel difference between the levels of the jet and propeller aircraft, when they sounded equally noisy. Boeing officials quizzed me about possible "bias" of the subjects used in his tests. I told them how carefully the subjects had been selected, and how meticulously the test questions had been drafted and vetted. The audience, comprising as it did the foremost experts in the field of aviation, knew how much different takeoffs can vary in noise level. A key question for them, therefore, was: "Are you making Boeing and Pan American meet the levels of the lowest-noise-level propeller-airplane takeoffs?" I showed them how much data we had on propeller aircraft operations at Idlewild and that the final selection of comparison numbers represented about 75 percent of the total operations, so there was no possibility of comparing the jet noise with the quietest of the propeller takeoffs. I went back to the hotel that night in utter exhaustion—but with the satisfaction of knowing that, for all its efforts, Boeing had failed to undermine our findings.

Apparently, nothing came of this meeting. As Tobin wrote in his July 23 letter to *Aviation Week:* "Throughout the day the Boeing engineers questioned

Dr. Beranek on the validity of this subjective differential of 15 decibels between hearers' response to a jet plane as compared to a piston engine plane. In our opinion, Dr. Beranek defended this theory and his figures very effectively, though [he admitted] future experience might indicate that the differential is several decibels lower than his present estimate of 15 decibels." The Boeing engineers, in effect, could envision no solution to the 707 noise problem. They did not want to install even heavier mufflers because this would force them to carry fewer passengers, to fly shorter distances (with less weight allowed for fuel), or to take off at Idlewild with partially filled tanks and then land in Boston to top them off. When told about the French and British takeoff procedure, they said they would speak with their pilots and bring Pan American into the mix.

After undertaking further research to improve its mufflers and adopting the new takeoff procedures, Boeing invited us to return to Seattle yet again to measure and record noise from the fully loaded 707. The Boeing executives told Tobin, Wiley, and my people that they would meet our airplane at the airport and take us to our hotel in company cars. By some quirk, however, unusually strong tailwinds from the East(!) pushed us to Seattle more than an hour ahead of schedule. With no one there to pick us up, we took cabs to our hotel. The chagrined executives apologized profusely that night at yet another elaborate dinner in our honor.

The next day, Boeing offered to take our party for a ride in the 707, at a high cruising altitude. The plane had only about 25 seats, and there was a bank of electronic equipment just behind the seats in racks bolted to the floor. About a half-dozen Boeing engineers came aboard to make instrument observations of the in-flight aerodynamics. I happened to be seated next to Boeing's president, William Allen. The flight was exhilarating—we took off gracefully and smoothly—free of the drone of propellers. After we reached an altitude of about 30,000 feet, the crew invited one of the pilots retained by the Port Authority to sit in the co-pilot's seat. All of a sudden, the plane began to shake violently, the engineers in our section sprinted in panic toward the cockpit, and Allen turned ashen white. Apparently, the guest pilot had pushed the button that lowered the landing gear. The Boeing pilot quickly retracted the gear and the turbulence ceased. We all breathed a sigh of relief, and several on board probably had felt for a moment as if they were approaching their maker.

In this series of tests, Boeing used the takeoff procedure initiated by the French Caravelle and British Comet 4 planes, that is, climbing steeply at full power and cutting back to normal takeoff thrust upon reaching an altitude of 1,100 feet. But the noise level measured in PNdB was still higher, by a few decibels, than that specified by the Port Authority. Tobin then decreed that, to meet the established noise standard, airplanes would have to turn away from residential areas as soon as they reached their leveling-off position.

Pan American officials strongly objected to a combination of a steep climb and a turn, arguing that such a procedure was unsafe. Pilots—not the Port Authority—would have to decide because Civil Aviation Board (CAB) rules stated that the pilot was responsible for aircraft safety. Pan Am tried to get the CAB to overrule the Port Authority, but it refused, observing that since the airport was not owned by the government, it would not meddle in its operations.

On July 23, the same day he wrote *Aviation Week,* Tobin also wrote to Western European airport operators that there was "serious doubt as to whether or not the fully loaded 707's [coast-to-coast for American or transatlantic for Pan American] can operate from New York International except under very severe operating limitations." Both London and Paris quickly took precautions against uncontrolled jet operations at their airports. Tobin's letter dismayed both airline and Boeing executives. In a related note, published a few months later, *Aviation Week* editorialized: "Port Authority's reaffirmation of its original 1951 ban of all jets not satisfying its noise level requirements served notice to manufacturers that they must make further efforts to suppress jet aircraft noise. Some credit certainly must go to the [Authority] for the fact that suppressors were designed and installed."

On July 25, Boeing ran its own subjective tests on the relative noisiness of jet and propeller-driven aircraft. Kryter and I were invited as observers, on condition that we would not disclose the results to the Port Authority or anybody outside of Boeing. We agreed to maintain secrecy, but only until the PNdB was officially adopted. The tests involved the same type of propeller and unmuffled jet airliners investigated in Kryter's laboratory studies. Boeing had access to a house in a Seattle suburb over which the planes flew. About twenty of their employees, armed with pencils and pads, sat in the open-windowed living room. Boeing may have suspected, or hoped, that people familiar with jet noise and loyal to Boeing might not be as sensitive as the subjects in Kryter's tests. Each person was asked to indicate the relative loudness of each overflight on a scale from 1 to 10. Engineers outside measured the noise levels during flyover. The two types of aircraft were flown at various heights and under various power conditions. Afterward, we were invited to sit with the engineers as they analyzed the data. When the results were tabulated, the engineer in charge turned to the Boeing managers and said: "BBN is right, the difference is 15 decibels." Boeing never publicly disclosed these confirmatory results.

A Final Decision on Noise Standards

Tobin now had to make a judgment about the noise level that the Port Authority would proclaim as the maximum that jet aircraft on takeoff could inflict on neighborhoods surrounding Idlewild. This involved a field test in August 1958. Tobin went to a home near the end of the runway in Howard Beach and sat on the porch. Several of us from Bolt Beranek & Newman went with him to take measurements. Whenever a propeller plane few over the house after takeoff, we would advise him of the maximum PNdB measured. After a time, he chimed in with the comment—simple, yet pointed—that a person owning a house near the airport should be able to sit on his porch and enjoy life. He was convinced, too, that this would be impossible if noise levels exceeded 112 PNdB, which thus became the standard—and has remained largely unchanged since.

At first, Tobin did not speak publicly of aircraft noise in PNdB. In a letter (available to the media) dated July 23, 1958, he mentioned only "Bolt Beranek & Newman's factor of subjective equivalence to the sound of a piston engine plane." *Aviation Week,* in an article of September 8, 1958, said that "the Port Authority will not publicly evaluate its noise test data." However, Tobin did quote one example of equivalent noisiness: "when measured by a standard sound level meter, the jet noise would have to be 15 decibels lower."

Then came a disturbing rumor. Some of the Port Authority staff reported that Pan American (perhaps in cooperation with Boeing) was planning to sue the Authority and BBN jointly, on grounds that there was insufficient basis for insisting on 15 decibels of noise reduction. The lawsuit, it was intimated, would be filed on the very day that the Port Authority was due to state official noise requirements for operations of the 707 out of Idlewild. It was also implied that Tobin had paid (bought off, essentially) BBN to confirm his desire that the aircraft be substantially quieter than Boeing had originally planned.

Pan American had inquired into the Port Authority's public record files and found that although Bolt Beranek & Newman had collected a "huge fee," not a single report prepared by BBN for the Port Authority on the noise of 707 jet aircraft could be found anywhere. That was true. The Authority had asked us to present all our data orally at meetings, generally held in New York, because of ongoing legal action in Newark. Reports received by the Port Authority were open to public inspection, and the plaintiffs in the Newark neighborhood's lawsuit against the Authority would have liked nothing better than written documentation of measured aircraft noise to buttress their own case. Many other communities around Idlewild, inspired by the Newark case, were mobilizing in anticipation of loud noise from projected jet aircraft operations. The Port Authority did not want our measurements and evaluations, particularly those with no mufflers, or with mufflers and no special takeoff requirements, to get into the hands of the general public. Unlike written reports, notes taken at our meetings were protected by executive privilege.

Following the lawsuit rumor, Tobin asked Bolt Beranek & Newman to assemble all data and to have them ready for inclusion in a detailed report. "When you have adequately organized the data, and I hope fairly soon," he added, "we will call a meeting of all our senior staff, including our lawyers, and have you make a presentation of your material. Then we will collectively make up our minds as to how far we want to go in pressing our noise requirements for the 707 Boeing jet operating out of Idlewild."

This meeting took place during the last week of August 1958. Afraid that it would become known to the media, all twenty-five of us, including Kryter and me, were flown in several helicopters from the top of the Port Authority building in Manhattan to the home of John Wiley, the Authority's director of aviation, in Connecticut. We landed, noisily, in his large backyard without permission from neighbors or the Federal Aviation Agency (FAA). The meeting ran for four hours. Kryter and I presented the measurements on propeller and 707 aircraft along with the psychoacoustic data that led to the creation of the PNdB method for rating aircraft "noisiness." Tobin told of his 112 PNdB ruling. We were rigorously questioned by Port Authority staff and lawyers, who then discussed the matter among themselves. They voted unanimously to adopt Tobin's noise-level limit for takeoffs at Idlewild, and agreed that BBN should produce its written reports by no later than October 4, 1958, the date the Authority had agreed to inform Pan American of all regulations required for 707 operations out of Idlewild. This gave us only a month to produce two mammoth reports—the other on the British Comet 4, to avoid the charge that the Port Authority was treating the British more favorably than the Americans. Riddled as it had become with problems, both technical and political, the whole process was now firmly within our grasp. We breathed a collective sigh of relief, and ended the night feasting on steak and lobster at a fine restaurant just down the road.

The Reports

I personally took on the job of managing work flow on the two reports. The next day I assembled all of BBN's technical staff who had been involved in the program and told them of the deadline and the material needed. The data and graphs were assembled by September 11. Our charge was to couch the text in terms that people living around Idlewild could understand as clearly as the Port Authority people and airline officials. Karl Kryter, Laymon Miller, and I divided the task and did as well as we could under pressure.

As we completed a particular section, I would fly with it to New York and sit down with Tobin, Lukens, Wiley, and legal counsel to polish the wording so that it could be understood by laymen and would not lend itself to confusion, misrepresentation, or legal action. No one at the Port Authority made any attempt to change our conclusions or to leave out data. A large number of charts were generated, and several of these—particularly the ones about the noise a muffled 707 engine would make as a function of temperature, altitude, and revolutions per minute (rpm)—were made available to Boeing engineers. One chart was modified in the final stretch to better illustrate noise produced at the higher rpm and altitudes.

Because there were no Xerox machines in those days, our drafts all had to be typed on mimeograph stencils, a cumbersome and now obsolete technology. A thin blue sheet of special composition attached to a heavy paper backing, the stencil and its backing were placed in a manual typewriter and the text typed onto them; the backing sheet was then removed and the blue stencil sheet placed on a perforated drum some 10 inches in diameter. As the drum rotated, the ink poured inside would pass through the stencil to create a duplicate copy on a sheet of paper. Drawings as well as text had to be incised onto mimeograph stencils. BBN draftspersons went to New York from Cambridge to do this; there were no computers in those days, no e-mailing of attachments. A thousand copies of each report would be required. The Port Authority claimed to have more mimeograph machines than any other organization in New York, so the printing job could be done in-house.

Two reports, each containing about 170 pages, were assembled in this laborious way. The main one was entitled *Studies of Noise Characteristics of the Boeing 707-120 Jet Airliner and of Large Conventional Propeller-Driven Airplanes;* the other had the same title, except that the name "Comet 4" was substituted for "Boeing 707-120."

The mimeographing was done by about 4 PM the day before the reports were due out. In an enormous room, eight squares were laid out, each about 20 feet on a side and made up of 3-foot-wide tables. The pages were placed in order on these squares. Ten pages took nearly 10 feet, so that 170 pages required 160 feet of tables, that is, two such squares for each of the two reports. Each page was in a pile 2 reams (or 1,000 sheets) high. There were no collating machines, and because time was so limited the Port Authority mobilized the help of some 20 Authority police officers to do this, all in full regalia with pistols and handcuffs bouncing on their belts. Each officer was to circle two squares and come up with a 170-page report, which was then placed in a box. For one report, this required 1,000 man-circles around two squares, about 50 for each officer—for two reports, twice as many.

Before midnight, all 2,000 copies were in boxes. With sirens wailing, several Port Authority police cars whisked them to the binders for stapling and covers. They were ready by 6 o'clock the next morning. Then a fleet of about a hundred taxicabs whisked them to the essential parties by 9 AM. No lawsuit ensued.

On October 4, both the *Wall Street Journal* and the *New York Times* summarized the process, its conclusion, the role of Bolt Beranek & Newman, and the Port Authority's establishment of a noise standard. The *Times* added that airline executives were anything but pleased: "Pan American says that if such operating conditions were to become permanent restrictions at Idlewild and other airports throughout the United States and the world, they would impose a severe, unjustifiable and, therefore, discriminatory handicap on all aircraft." Later in October, the *Times* reported that between October 4 and 13, one-fifth of the trial jet takeoffs at Idlewild by BOAC and Pan American— 5 out of 25—had successfully followed the climb and turn rules. John Wiley sent the pilots letters of congratulation. The *Times* also cited a Port Authority spokesman's comment that sound meters and cameras had been placed in nearby communities to record takeoffs. Microphone, camera, and recording equipment at 2.5 miles from start to takeoff roll, at the edge of Howard Beach, were supplied by BBN.

The Jet Age Begins

Starting late in the afternoon of October 26, 1958, a Pan Am Boeing 707-120, with a full complement of passengers, flew from Idlewild to Heathrow. That evening, a BOAC Comet 4, fully loaded, flew from Heathrow to Idlewild. These two events marked the entry of the United States into the age of commercial jet travel. There were no complaints about noise.

By the spring of 1959, the Port Authority had issued a new rule that noise over communities from jet aircraft must not exceed 112 PNdB as measured outdoors on the ground—a level requiring 1,200 feet in altitude and 8,000 pounds of thrust for each engine. Under the original rules, the Authority specified certain flight procedures that pilots were to follow after takeoff, including power settings and turns. With the 112 PNdB rule over populated areas, however, the method of flying was now up to the carrier. Compliance with the antinoise rules stood at 70 percent in July 1959, improving to 92 percent a year later.

Airlines and airplane manufacturers in the United States fought the regulations for nearly a decade. Bolt Beranek & Newman (Beranek in particular) was vilified by the industry. When I was elected to the National Academy of Engineering in 1966, an aircraft engineer elected at the same time openly expressed his distaste for BBN and refused to talk to me. My reputation had preceded me, so to speak.

In the mid-1960s, the FAA made a second study around Idlewild Airport (called John F. Kennedy Airport by this time) using personnel from NASA. Their results confirmed ours. BBN took measurements in the neighborhoods around JFK in late 1965, showing that because the Port Authority monitored noise at only one location, airline pilots were cutting power—tactically—as they flew over that location but then reverting to full power immediately afterward. The NASA report confirmed this. To counteract this ploy, the Authority added measuring positions. Industry officials thus continued to view BBN as a thorn in their side.

In November 1966, the FAA sponsored a meeting of airport operators, airline executives, airplane manufacturers, and government regulators at Lancaster House in London: the "International Conference on Aircraft Noise." The BBN people were not invited. When asked about this, the FAA admitted that Beranek was not welcome there. Undeterred, I went to London and gained admission simply by walking in—probably the only time I ever gate-crashed a function. Many attendees spoke about the noise problem and the use of PNdB to evaluate it. I joined in the discussion.

A professor from the Technical University in Copenhagen, Fritz Ingerslev, moved that PNdB be officially adopted as the international standard for measuring aircraft noise. Frank Kolk, a vice president of American Airlines, urged that the motion be approved. He argued that it was the best known measure and the result of careful and substantiated research, and that its international adoption would help put a stop to ongoing controversy and dissension. The motion carried.

Almost exactly four decades later, in 2006, Boeing and General Electric proudly announced that the new Boeing 747-8 had measured noise levels that were 10 decibels below those of the 747-400 and 10 decibels below International Civil Aviation Organizations (ICAO) standards. "These results demonstrate," Boeing declared, "our effort to design the 747-8 with the community and environment in mind." All in all, a remarkable and welcome change of heart over the years.

The Port Authority's vigilance and federal regulations motivated the ICAO standards. Credit goes to Austin Tobin for financing the BNN study, for accepting its results, and for setting and enforcing limits of "noisiness" over the intense objections of industry and government. Bolt Beranek & Newman helped balance conflicting commercial and community interests, averting, for example, the specter of entire families parading onto runways as human shields to prevent flights from either landing or taking off. The initial noise regulations led to a speedup in the development of by-pass jet engines and, a decade later, to the FAA's promulgating the FAR 36 noise rules for the design of new jet aircraft, and to hush kits for existing jet aircraft. Engine and aircraft manufacturers deserve credit for ultimately accepting noise standards, and for exercising due diligence in the design of quieter jet engines.

First day at school, Tipton, Iowa, with
Collie, 1919.

Horse-drawn school bus, Iowa Consolidated Schools, 1920.

Holding reins of
Pony, 1923.

Phyllis and Leo, wedding day,
September 6, 1941.

Director of Electro-Acoustic Laboratory,
Harvard, 1943.

Four generations of Beraneks; Edward, Anna, Leo, Thomas, and James, 1956.

Last gig as a trap drummer, 1950.

Skiing, 1975.

With Dick Bolt and Bob Newman, 1951.

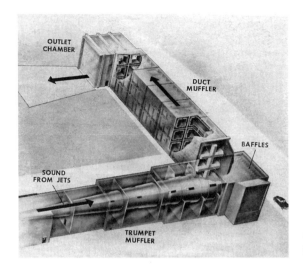

World's largest muffler,
NASA, Cleveland, 1951.

With Austin Tobin and John Wiley of Port of New York Authority, measuring Boeing
707 noise, Boeing Field, Seattle, 1958.

With Leonard Bernstein,
New York, 1959.

Preparing to fly in Piasecki helicopter, with Laymon Miller and Piasecki engineers, wearing parachutes, 1960.

With Jordan Baruch, Dick Bolt, Sam Labate, and Bob Newman on initial public offering day at Bolt Beranek & Newman, Inc., June 27, 1961.

With Max Abramowitz, Thaddeus Crapster, and Walfredo Toscani, in architect's office, discussing Philharmonic Hall, 1961.

With son Thomas, witnessing first on-air transmission, WCVB-TV, 4 AM, September 10, 1971.

Behind the monotony of legal language is an exciting story of intrigue, surprise, big names, vast power, struggles for survival among corporations . . .

BBI PRESIDENT LEO L. BERANEK AND WCVB-TV'S STATION IDENTIFICATION IN DOUBLE EXPOSURE (Globe photo by Joseph Dennehy)

Channel 5: longest license battle ends

Across-the-page spread in the *Boston Globe* on the occasion of FCC granting on-air go-ahead to WCVB-TV, January 23, 1972.

With Bill Poorvu and Bob Bennett, builders of WCVB-TV, 1972.

Leo and Gabriella, wedding day, August 10, 1985.

Conducting *Stars and Stripes Forever*, Boston Pops Orchestra in Symphony Hall at a regular concert, June 27, 1988.

With George W. Bush, receiving National Medal of Science, White House, September 2, 2003.

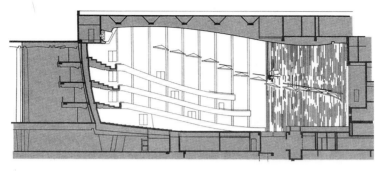

Architect's drawing, Philharmonic Hall, New York, 1972.

With Japanese colleagues, engineer Takayuki Hidaka and architect Takahiko Yanagisawa, 1994.

Tokyo Opera City Concert Hall, 1997.

7 Music, Acoustics, and Architecture

It was a sultry summer day in 1950 at Bolt Beranek & Newman—quiet, too, with several staffers away on vacation. The thought of cold lemonade followed by a siesta kept tempting me away from BBN's unair-conditioned offices at 16 Eliot Street. I was waiting for Zeev Rechter, an architect from Tel Aviv, whose flight had been delayed several hours. Around 4 PM, in walked an older gentleman with a distinguished crop of gray hair. By the way he fixed his eyes on me, I knew straight off he had come on serious business.

I listened as Rechter laid out his ambitious project: a concert hall for the capital of a country founded just two years earlier after decades of debate, struggle, and warfare. He said that Wallace Harrison, the lead architect for the United Nations Headquarters in New York, had recommended us on the basis of our successful work on the United Nations General Assembly building. I told Rechter we had almost no experience with concert halls, although, at the time, we were working with Eero Saarinen on the acoustics for Kresge Auditorium at MIT and with architects for several universities considering music buildings. I recommended a visit to Boston's Symphony Hall—just to get a sense of the layout, no concerts were scheduled in the summer—but Rechter was not looking for a classic shoebox-shaped hall, so common throughout Europe. I told him about the Hill Auditorium in Ann Arbor, Michigan, and the Kleinhans Hall in Buffalo. I had never visited either of these, but I laid out what I had learned about them from the acoustics literature—along with my analysis of the pros and cons of different halls. We discussed fees and he said he would be in touch.

In September, a letter arrived offering BBN the acoustical consulting contract and asking that we come to Tel Aviv immediately. I took the assignment and went over to meet with the consortium architects. Zeev Rechter along with another Tel Aviv architect, D. Karmi, and Zeev's son, Jacob Rechter. They reiterated that both the architects and the building committee wanted, not a shoebox-shaped hall like those in Europe and Boston, but something more along the lines of Kleinhans Hall, which came highly recommended by Serge Koussevitzky, celebrated conductor of the Boston Symphony Orchestra. The Tel Aviv plan called for some 2,700 good-sized seats, best accommodated, the principals felt, without sacrificing all semblance of intimacy, in a shorter, broader, fan-shaped space.

To investigate the Kleinhans Hall, Dick Bolt and John Kessler attended a symphony concert there in the late summer of 1950 and asked audience members, for the most part not regular symphony subscribers, whose opinions might have been less complimentary, to complete a questionnaire. Almost all responded that the loudness, tone quality, and reverberation were about right, and the acoustical definition was excellent. BBN had not taken negative media comments seriously because the Kleinhans acoustical consultant had quoted a reverberation time—the time it takes a loud sound to die away to inaudibility—of 1.8 to 2.0 seconds. Some years later, BBN measured the middle-frequency reverberation time in the fully occupied hall, using orchestral stop chords as the source of sound, and found it to be 1.35 seconds, far less than 1.8 to 2.0 seconds. The acoustical consultant's figure had been calculated in advance and not confirmed later.

In any event, we based our calculations for the Tel Aviv hall heavily on information published by the acoustical consultants for the Royal Festival Hall in London, which opened in 1951—one of the first halls completed anywhere after World War II. The London hall's consultants expected a reverberation time of 1.7 seconds at 500 hertz with a hall volume of 210 cubic feet per seat. For the Tel Aviv hall, the architects went along with our recommended volume of 270 cubic feet per seat, and, because reverberation time rises with increased volume, we expected that this would safely result in value of 1.9 seconds, which became our prediction for the hall. When the Royal Festival Hall had opened in 1951, however, the reverberation time measured only 1.45 seconds. The London consultants attributed the shorter time to "an unexpectedly large ceiling absorption." We made sure the ceiling structure in Tel Aviv was not excessively absorbent. Although this explanation seemed plausible, I would discover—too late, unfortunately, for application to the Tel Aviv hall—that it was deeply flawed.

Acoustical Detective Work

In the late spring of 1957, along with other acoustical consultants, I was invited to Edmonton, Alberta, to evaluate a new hall, the Jubilee Auditorium. After listening to a concert, most of us felt that the reverberation time was too short. We were told that the predicted time had been 1.8 seconds, whereas the actual time, measured a few days before our visit, was only 1.4 seconds. I then recalled that Wallace Sabine, acoustical consultant for Boston's Symphony Hall decades earlier, had also predicted a longer than actual reverberation time, in this case 2.3 versus 1.9 seconds.

Here was the fourth concert hall (Boston, London, Buffalo, and now Edmonton) in which the reverberation time measured after construction was much shorter than the time calculated in advance. I set out to discover why. As everyone knows, when a carpet is added to a bare living room floor, the reverberation is diminished, that is, the carpet absorbs sound. But what is it that determines the amount of sound absorption by a carpet? Is it the number of tufts or the overall area of the carpet? An audience in a concert hall also absorbs sound. Does the level of absorption depend on the number of people in the audience—the number of tufts in the carpet—or on the overall area of where they are seated?

I went to the literature and found the reverberation times for 40 concert halls. I then related these measurements, first to the volume per seat and then to the volume per 1,000 square feet of audience. Strikingly, the correlation between the RT's and the number of audience members was small. In contrast, there was a high correlation between reverberation times and audience areas. Thus, it is not the number of tufts in the "carpet" but rather the area of the "carpet" that determines its sound absorption.

Consultants for the Royal Festival Hall had wrongly assumed that audience absorption was proportional to the number of people, and they worked on the notion that the amount per person was about the same as that measured in the Amsterdam Concertgebouw. But the number of people per 1,000 square feet in the Royal Festival Hall is only 176, not 224, as it is for the Concertgebouw. Hence, as my finding shows, each person in the London hall absorbs far more sound than they assumed, which explains why the reverberation time there is less than they had calculated in advance. Call it a "Eureka!" moment for me, the proverbial lightbulb over the head, whatever you like.

I recalculated for the Tel Aviv auditorium and found that the reverberation time at middle frequencies would most likely be about 1.5 seconds, significantly shorter than the 1.9 seconds we had first predicted. With the opening fast approaching (1957), I headed off to Tel Aviv to give the architects the devastating news. It was obviously too late to increase the hall's reverberation time by raising the ceiling to boost the hall's volume. The actual time measured 1.55 seconds. My findings on audience absorption in concert halls, first presented at a meeting of the Acoustical Society of America in May 1958 and published in *JASA* in June 1960, proved to be of major importance to acousticians.

Fresh Momentum

A basic tenet of concert hall acoustics is that every hall that looks different sounds different, just as a square violin will not sound like a Stradivarius. Some hall shapes can be augmented with reflecting surfaces, hanging panels, and absorbing surfaces, so that their acoustics approach that of the best halls. Other shapes almost defy correction. One example of a difficult design is MIT's Kresge Auditorium. Designed by Eero Saarinen, it is shaped like a half grapefruit standing on short vertical sidewalls. Engaged as acoustical consultants, Bolt Beranek & Newman chose to mitigate the extreme focusing effect of the dome by adding areas of hanging panels. The sound in this small hall (1,238 seats), with its low panels over the stage, basic hemispherical shape, and reverberation time of only 1.5 seconds, was understandably inferior to that in Boston's Symphony Hall, with its high stage ceiling, no panels over the stage, shoebox shape, and reverberation time of 1.9 seconds. After an opening concert by the Boston Symphony Orchestra in 1955, the New York music critics, instead of praising the hall for successfully minimizing the impact of impossible acoustical defects, dubbed Kresge an "acoustical failure," an opinion not at all shared by the *Boston Globe,* which observed that a string quartet was "perfectly played and superbly realized through the marvelous acoustics." Also commenting on the string quartet, the critic for the *Jewish Advocate* raved that "the quality of tone in the Kresge Auditorium was so beautiful that I for one was grateful to hear the instrumental element so clearly." And later, following a piano recital, the *Christian Science Monitor* singled out for special mention "the acoustic excellence of the Kresge Auditorium, itself the effective product of plan and know-how."

Another almost impossibly shaped auditorium was the Aula Magna Concert Hall, on the campus of the Universidad Central de Venezuela in Caracas, designed by architect Carlos R. Villanueva, and constructed by engineers Santiago Briceño-Ecker and Daniel Ellenberg. The broad, fan-shaped hall has a domed ceiling and a rear wall that forms a sector of a circle with its center of curvature at the rear of the stage. In 1951, Bolt Beranek & Newman was commissioned to advise on the acoustics. The hall's domed ceiling and the circular rear wall presented acoustical problems similar to those of the Kresge Auditorium in Cambridge: focused echoes, dead spots, and a general lack of uniformity in sound distribution. After rejecting our first recommendation—to greatly modify the interior shape of the hall—the university and the architect accepted our "Plan B": to install a number of large sound-reflecting panels, whose total area amounted to about 70 percent that of the ceiling, hung below the domed ceiling and on the side walls. Hoping for more visually appealing shapes than our proposed rectangles, the architect and the university contacted Alexander Calder, a Paris-based American sculptor known for his mobiles.

Calder became part of an unusual and rewarding collaboration between sculptor, architect, engineers, and acoustical consultants. The result is an exciting array of "stabiles," beautiful in both form and color, suspended from the ceiling and standing away from the side walls. No photograph can do the building justice. You have to be inside the hall—inside the sculpture—to feel its rhythm and color. Operating in 1954, the Aula Magna Concert Hall seats 2,660 people and is used for college events and conventions as well as for musical events. The hall's short, reverberation time, only about 1.4 seconds at middle frequencies for full audience, is a drawback for symphonic music. The Aula Magna is best for piano, chamber music, and modern music in which clarity and brilliant string tone are essential for conveying complex tonal relationships. But even so, after conducting the New York Philharmonic there in 1957, Leonard Bernstein gushed: "I wish I could take the stage part of the hall back to New York for the Philharmonic to use." In those days, the New York Philharmonic performed in Carnegie Hall.

Bolt Beranek & Newman was often asked to advise on the acoustics of multipurpose halls. My first large-sized hall of this type was the Binyanei Ha'Oomah Congress Hall in Jerusalem, designed by architects Rechter, Zarhy, and Rechter of Tel Aviv. Seating 3,200, the hall opened in May 1960. Fortunately, the Mann Auditorium in Tel Aviv had opened three years earlier;

because two of the architects were involved in this new project, the pitfalls of Tel Aviv were avoided. The reverberation time at middle frequencies with hall fully occupied is 1.75 seconds, near optimum for symphonic music. Reviews by conductors and music critics alike have been uniformly enthusiastic. "A conductor expects a hall to allow him to hear all instruments in correct balance," the Detroit Symphony Orchestra's Paul Paray wrote. "The new hall does this, and in addition, gives each instrument a beautiful resonance. I am perfectly satisfied."

Another BBN project, the Grace Raney Rogers Auditorium in the Metropolitan Museum of Art in New York, opened in 1954 to universal praise from musicians. Leopold Stokowski found it "a joy to make music in a place where the acoustical conditions have been created with such skill and deep understanding of what music requires in an enclosed space." But because of its size, just 708 seats, the auditorium is best for chamber music or solo recitals; symphonic music generally comes across as too loud. Other halls in BBN's earlier days included the Benjamin Franklin Kongresshalle in Berlin, specializing in chamber music, the Dartmouth College Arts Center Hall in Hanover, New Hampshire, the Heinz Auditorium in Pittsburgh, and the Clowes Hall in Indianapolis—all considered acoustical successes.

Opportunity of a Lifetime: The Lincoln Center Project

BBN's reputation had emerged almost overnight from our work on the United Nations Headquarters. By the mid-1950s, we were acoustical consultants of choice. It was therefore no great surprise when, in 1955, Wallace Harrison, chief architect for the UN project and a partner in the firm of Harrison and Abramowitz, called to invite us to a meeting in New York to discuss the design of a new set of buildings to be called "The Lincoln Center for the Performing Arts."

The push for Lincoln Center had begun with the recommendation by a municipal committee that a slum area on Manhattan's West Side be redeveloped. Meanwhile, the Metropolitan Opera Association was looking to acquire a site for a new opera house, and the private owners of Carnegie Hall had notified the New York Philharmonic that their lease would not be renewed and the building demolished. A city-sponsored planning committee was appointed, the land cleared and the tenants moved out. The board of the newly formed Lincoln Center Corporation selected Wallace Harrison as architect for the Metropolitan Opera House and his partner, Max Abramowitz, as architect for the Philharmonic Hall. Abramowitz then chose us as acoustical consultants for the Philharmonic Hall, along with Hugh Bagenal, the British acoustician who had helped out with the Royal Festival Hall in London. From the beginning, we were assured that "optimum acoustical properties were paramount."

Seeking Perfection with Field Surveys

To determine, in the words of the Lincoln Center committee, "the relationship between acoustical properties and size and design of auditoria," BBN decided to undertake a study of existing concert halls and opera houses, primarily in Europe, and to interview the world's leading conductors and music critics. I agreed to visit the better-known halls and gather opinions from conductors and music critics. I also got in touch with leading acoustical consultants worldwide, all of whom agreed to share their findings with me.

I visited one country at a time from 1955 to 1961. In Great Britain, Tom Somerville of the BBC and acoustical consultant William Allen helped me acquire seats for concerts in the main London and Scotland halls. I found it harder to get tickets for events in outlying areas. I came up with a strategy to beat the odds. On hearing that a certain concert was sold out, I would learn the name of the general manager of the local hall, then go around to the stage door and ask to see him. Not long after telling the manager I was the acoustical consultant for the Lincoln Center Concert Hall to be built in New York and that hearing a concert in his hall would be of enormous help to me, there would invariably be a ticket waiting for me at the box office.

The halls were widely scattered and presented concerts on only a few nights each week, so I zigzagged my way by train from London to Edinburgh, to Manchester, to Bristol, to Glasgow, to Liverpool, and back to London. As I worked my way north, then south, staying in small hotels, I have vivid memories of sleeping in beds with old-fashioned hot-water bottles to warm my feet. My 1962 book *Music Acoustics and Architecture* relates my findings hall by hall in considerable detail. Thus, for example, the halls in England: "Colston Hall [in Bristol] is very good, almost excellent"; "Usher Hall [in Edinburgh] ranks below Colston Hall and St. Andrews [in Glasgow] is very fine and could easily be one of the great halls of the world [except for a] large amount of thin wood that reduces the reverberation time at low frequencies." "I have never heard an orchestra or a piano sound as weak [as in Royal Albert Hall, London]. But [when] the orchestra played Tchaikovsky's *1812 Overture*[, t]he audience was left breathless and tingling. It is for these moments of ecstasy that the Albert Hall continues to exist." "The definition [in Royal Festival Hall, London] is excellent and the hall is very good for piano, chamber music, and modern music. On the main floor the reverberation is hardly noticeable—there is little liveness. My feeling is that the lack of bass is the most serious problem."

I also traveled to halls in Austria, Germany, Denmark, Finland, France, the Netherlands, and Sweden, and during month-long skiing vacations in Switzerland nearly every winter, I surveyed halls in Zurich and Basel. I taught for three months in the summer of 1949 (their winter) at the University of Buenos Aires, where I attended performances at the famous Teatro Colón and assessed its acoustical features. This was a period, too, when BBN consulted on several halls in the United States; I often took weekend trips to attend concerts at those venues.

As part of my preparation for the Lincoln Center work, I interviewed musicians, particularly conductors. Meeting with Herbert von Karajan in Vienna, I found him more accommodating than expected—nothing like his reputation for aloofness among all but his closest friends and colleagues. I spent a part of each of four days with him, September 13 to 16, 1959. At the time, von Karajan was arguably the best-known conductor in the world, certainly the most recorded. He had a crushing schedule leading the Vienna State Opera, Vienna Philharmonic Orchestra, and Berlin Philharmonic Orchestra—all in overlapping time frames. He was also artistic director of the Salzburg (summer) Festival and a regular conductor at La Scala in Milan. A joke went around that once, after a daytime rehearsal in Salzburg, he hailed a taxi and when the driver asked him where he wanted to go, he replied: "Anywhere, I have business everywhere."

Von Karajan struck me as brilliant, dynamic, outspoken, sure of himself, and utterly devoted to music. He was candid and direct in his opinions about the halls he knew. After our sandwich lunch in the green room of the Vienna State Opera the first day, he got me a ticket for that evening's opera to sit with his wife in his box while he conducted. Another day, he had his Mercedes bring me to the Sofiensaal, where he was recording the opera *Tosca*, after which we went to a restaurant and chatted some more. Most unusual of all, I attended a rehearsal of the Vienna Philharmonic in the Musikvereinssaal on Tuesday, September 15. Von Karajan normally allowed only one person to attend rehearsals, a mysterious lady dressed in black, with a black shawl draped over her shoulders. I was honored to be a second listener that day, and took acoustical notes on the hall. Afterward, we lunched and talked once more. I couldn't help wondering how this warm, helpful man could possibly have earned the epithets "distant," "authoritative," "haughty," and "egotistical."

Although von Karajan felt that Vienna's Musikverein was one of the best halls in Europe, he liked Boston's Symphony Hall better because sound there did not come across so crushingly loud. I was surprised when he spoke favorably of London's Royal Festival Hall, which has never received high ratings for its acoustics. He found the sound there clear and faithfully projected to the audience, although the hall's reverberation was too low. He rated Carnegie Hall as only "good" to "fair"; he considered Chicago's Orchestra Hall perfect *without* an audience (he had made a recording there). He praised the acoustics of the Garnier Opera House in Paris, "whose ceiling," he said, "reflects back a beautiful tone, and where crescendos can be conveyed with precise nuance." He especially liked the sound in the Stadt-Casino in Basil, Switzerland. As for opera houses, he thought the Vienna State Opera the best, but he was also fond of conducting in La Scala—partly for its acoustics, but also for its animated audiences. All in all, he was remarkably helpful and informative.

Next I interviewed Dimitri Mitropoulos in his Vienna apartment. His ideal concert hall design was based on Greek theater; halls based on other concepts were hit or miss. He rated a half dozen or so halls, after which I made a graph to see whether his impressions correlated with certain defined variables, particularly spatial volume. It turned out that his best-liked halls had volumes of about 300,000 cubic feet (Basel), and his least-liked about 1 million cubic feet (Bloomington, Indiana, and Rochester, New York). I made a similar graph after interviewing Eric Leinsdorf, with almost identical results.

Mitropoulos gave me permission to sit in the pit of the Vienna State Opera the next time he conducted an opera. I was told to wear a dark suit with a white shirt and special gray tie in order to blend in with the orchestra. A chair was set aside for me at the second desk in the second violin section next to the pit railing, on the right side of the conductor, so that I would not be visible to the audience. Imagine my heart beating as the lights dimmed. We waited and waited, but Mitropoulos did not appear—I saw him once looking through a small window in the door at my end of the pit. After a few minutes, a man in a green uniform walked out on stage and announced that the performance was delayed because an orchestra member had fallen ill. Already nervous, I now became alarmed. Had there been a complaint about me? I could see no sick man in the pit and the violinist next to me couldn't either. Ten minutes later, much to my relief, the lights again dimmed, Mitropoulos took the podium, and the performance began. I learned afterward that the French horns at the other end of the pit had been given the wrong parts for the opera, and that the librarian had had trouble finding the right ones, which had been accidentally folded into the score set for yet another opera. From my perch, and for the first time in my life, I heard the sounds of an orchestra from within the walls of its pit 5 of 6 feet below the level of the first row of audience seats. Each instrument seemed nearer and louder than ever before, and the conductor's every gesture conveyed shades of dynamics with subtlety yet precision.

In the late fall of 1959, I visited with the great conductor Leopold Stokowski in New York. The best halls, in his view, were the Musikverein in Vienna, the Academy of Music in Philadelphia, and the Philharmonia in Saint Petersburg; the Musikverein outranked both Boston's Symphony Hall and Amsterdam's Concertgebouw. He felt that the bass tones in Carnegie Hall should be more prominent and the highest violin tones somewhat less so. We talked about the design features I was planning to present to the architects for Lincoln Center, and we pretty much agreed that my best course of action was to copy Boston's Symphony Hall but with fewer seats. Sometime later, when he came to Boston as guest conductor, I took him in my car on a tour of the city. He talked freely about his experiments with alternative seatings for certain instrumental sections in the Philadelphia Symphony Orchestra, and how he had overcome the lack of reverberation in the Academy of Music by having the string players extend their bowing motions.

To get an appointment with Antonio Ghiringhelli, the top man at the La Scala Opera House in Milan and one of Italy's icons, was nearly impossible, but thanks to my Lincoln Center connection, I was granted an interview on February 10, 1960. Maintaining that any new opera house should use Vienna as a model, Ghiringhelli told me how experiments at La Scala had led to the conclusion that the overhang to the pit should be limited to one meter. We talked about these and other matters for an hour, after which, he arranged for two "Gala" tickets—that same night for seats on the main floor of La Scala. The concierge in our hotel told me that "Gala" meant black tie for men and evening dress for women. The most formal clothes I had with me were a dark suit, white shirt, and black bow tie.

The evening was cold, so Phyllis wore a fur coat and I a topcoat. When we arrived at the opera house, she was waved in but I was stopped. The usher explained that, for a Gala, I must wear a white silk scarf and that since I was wearing a plaid woolen one, he had no choice but to turn me away from the main floor seating. Even though I would be checking my coat and scarf, the rule applied to attire in the outer vestibule as well. The usher then handed me a ticket to a box on the uppermost tier, tucked away at the side adjacent to the proscenium, hardly the best place to see or hear. By a fortunate coincidence, however, the architect who had rebuilt La Scala after its near-total destruction in World War II was sitting in the same box. I introduced myself, and we spoke about my project and the La Scala reconstruction. He had insisted that the ceiling be supported from the beams above by leather thongs, just as they had been in the original design. Imagine my surprise a few months later to read that an hour or so after the end of one performance, the ceiling had fallen to the floor—apparently, leather thongs were more robust in the eighteenth century. At intermission, I disappeared into the crowd on the main floor and sat with my wife for the rest of the performance.

Among the other conductors and musicians I interviewed were Eugene Ormandy, Fritz Reiner, Isaac Stern, Pierre Monteux, Charles Munch, Herman Scherchen, William Steinberg, Tauno Hannikainen, Igor Markevich, Erich Leinsdorf, Brenda Lewis, Izler Solomon, Stanislaw Skrowaczewski, and Bruno Walter. (My interview notes can be found among my papers in the archives at MIT.) I also interviewed a wide range of music critics, from Paul Hume of the *Washington Post,* Harold Taubman of the *New York Times,* and Alfred Frankenstein of the *San Francisco Chronicle* to Frank Howes of the London *Times* and Colin Mason of the Manchester *Guardian.*

Tackling the Tanglewood Music Shed

About this time, I became intimately involved in a related BBN project—acoustical improvements to the summer home of the Boston Symphony Orchestra in the hills of western Massachusetts. Designed by Eliel Saarinen, father of the architect of MIT's Kresge Auditorium, the Tanglewood Music Shed (now called the "Serge Koussevitzky Music Shed") had been dedicated in 1938. The hall's original wide, fan shape survives, but, for budgetary reasons, Saarinen's elegant design was discarded in favor of a purely functional one. When the building committee had informed him that the hall needed to be drastically redesigned, he had withdrawn, saying: "If I do that, you will only have a shed, not a concert hall." Hence the hall's sobriquet.

With a construction budget of $100,000, amazingly little for a hall accommodating 6,000 listeners (today 5,000 because of larger seats), an engineer in Stockbridge, Massachusetts, Joseph Franz, took over the project. The first saving Franz made was to recycle a tent-sheltered stage enclosure used for previous Tanglewood concerts by moving it to the front of the new Shed. He then had seats mounted on the earth in the audience area, which had been graded to the desired slope, a feature that remains to this day. He reduced costs for the roof by leaving the internal structural girders exposed to view. The lower half of the rear and side walls were, and remain, open to the outdoors; as many as 10,000 people can be seated on the lawn outside. When the Shed opened, music critics agreed that the sound was muddy, too reverberant, and with little definition—nothing projected clearly. They also agreed that, of all performers, singers and pianists fared the worst.

More than a decade later, in 1954, the Boston Symphony Orchestra (BSO) hired Bolt Beranek & Newman to carry out an acoustical study of the Shed. Again and again over several years, Bolt, Newman, and I met with Henry Cabot, BSO president, and Thomas D. "Tod" Perry, general manager, to discuss prospective designs to better the sound. None appealed to them until we came up with the present one: a specially shaped structure with zigzag sidewalls and with an acoustical canopy made from triangular panels of varying size at a height of about 25 feet above the stage and extending past it over the first 13 rows of seats. Because the panels touch each other tip to tip, part of the sound passes through the canopy to create the desired reverberation, and part is reflected directly to the audience to produce the desired clarity of sound. A series of sloping panels on the upper wall at the rear of the hall reduces echoes and further equalizes the sound carried out to the audience. Meanwhile, the Boston Symphony Orchestra engaged architect Warren Platner of Eero Saarinen and Associates to beautify our design and to integrate the lighting.

The inaugural concert for the newly refurbished Shed took place in July 1959. I sat with the *New York Times* music critic, Harold Taubman, who was most complimentary. After intermission, we walked out to hear the sound on the lawn and agreed that it had never been as well projected or as clear. "Most people, on first hearing," a Montreal critic wrote, "are convinced that an elaborate system of electrical amplification has been installed, but in actual fact there is not a single speaker in the hall." Many people felt that the strings, and especially the violins, came across in the new Shed with the brilliance and clarity of the best concert halls. "One can genuinely enjoy Bach there," Eugene Ormandy remarked. "It does not sound like a 6,000-seat hall." We were elated by the result, and we believed that a similar canopy would make sense for the Lincoln Center home of the New York Philharmonic.

Tell It in a Book

By 1960, Bolt Beranek & Newman's experience on hall acoustics for musical performance exceeded that of any other firm or consultant. We had studied halls that differed widely in shape, occupancy, appearance, architectural details, and sound quality. We adopted as our basic tenet: "Every hall that looks different sounds different." We came to the conclusion, both from our interviews and from audience reaction to the halls we worked on, that the acoustically best venue for modern-day orchestral repertoire is a shoebox-shaped hall in which the audience size does not exceed 2,400. A variety of architectural details—some unobserved or even undetectable by an audience—is critical, and must be framed and adapted to the size and shape of each architect's unique hall. With this in mind, I decided to incorporate my findings on 54 halls into a book, *Music, Acoustics, and Architecture,* published by Wiley in 1962.

Boston's Symphony Hall: A Model Precedent

A question often asked at the time, by people at Lincoln Center and elsewhere was: "Why did the acoustics of the Boston, Vienna, and Amsterdam halls come out so well when they were built so long ago, in 1900 or before?" To answer, I studied what had happened in the case of Boston's Symphony Hall. The story began with an assistant professor at Harvard, Wallace Clement Sabine, who took on the job of improving the acoustics of an auditorium in Harvard's former Fogg Art Museum. In the summer of 1898, when Sabine was just 30 years old, his recommendations were carried out and he noted in a published paper: "The result has been pronounced entirely satisfactory." Harvard's president, Charles Eliot, was so impressed that during a chance encounter with Henry Lee Higginson, chairman of the building committee for the proposed Symphony Hall, he suggested that the Boston Symphony Orchestra might benefit from Sabine's knowledge of acoustics. Higginson was intrigued by the idea. When Eliot contacted Sabine, however, he said he would not consider meeting with Higginson until he had made sense of the large amount of reverberation data he had accumulated on 11 other halls at Harvard during the course of the Fogg study.

Unmarried and living alone with his mother, Sabine isolated himself in an upstairs room of their house in Cambridge. For two weeks, he feverishly studied and restudied the fruits of his labor over the previous three years. Suddenly, he rushed downstairs and announced to his mother, "I have found it at last!" "He gets a perfect hyperbolic curve," she wrote in a letter to her daughter. "This makes everything definite. His whole face smiles, although he is very tired." The next day, October 30, 1898, Sabine wrote to Eliot offering to act as acoustical advisor to Henry Lee Higginson: "This discovery opens up a wide field. . . . One only needs to collect further data [on the absorbing power of various building materials] to predict the character of any room that may be planned, at least as respects reverberation."

Sabine met with Higginson about a week later. Higginson was so impressed by the young scientist's incisive approach to acoustical problems that he took careful notes and kept them for future reference. No fee was offered to Sabine, he was expected only to give some off-the-cuff advice. Sabine took this prospect seriously and proceeded immediately to measure, in his Harvard laboratory, what he recognized as critical needs: the sound-absorbing power of acoustical materials such as wood, draperies, and plaster on lath. He also determined the sound absorption of a lecture hall audience in the Jefferson Physical Laboratory. According to his "hyperbolic equation," the reverberation time in a room is directly proportional to the volume of the room divided by the total absorption provided by audience and the surfaces in the room.

Higginson had decided some years before that the new hall should accommodate about 2,500 people. He engaged Charles McKim of McKim, Mead & White, celebrated New York architects, to put together a design. McKim came up with a hall that resembled an outdoor Greek theater covered by a roof. Higginson carried photographs and drawings of McKim's model to Europe, where he consulted with leading conductors and music critics (much as I would six decades later on behalf of Lincoln Center). The consensus was that no hall of such a design had ever been successful acoustically. Instead, they recommended the Gewandhaus Concert Hall in Leipzig, destroyed in World War II, as a model.

In mid-November 1898, Higginson asked McKim to design a hall with the shoebox shape of Boston's old Music Hall (from which the Boston Symphony

Orchestra was departing) and the Gewandhaus. Although unhappy at the prospect, McKim promptly sent back sketches of three possible designs. One of these simulated the Gewandhaus, but with all dimensions increased by 30 percent yielding a seating area for 2,600 persons. Without telling McKim, Higginson gave this expanded design to Sabine, who calculated that the reverberation time in such a hall would be 31 percent greater than in the Gewandhaus, a very noticeable difference. Higginson told McKim that, although the general plan for a hall modeled on an expanded Gewandhaus was a move in the right direction, for visual reasons, the width of the hall should be less than the 77 feet of the Music Hall, and the ceiling height should be lowered to reduce hall volume by about 30 percent. Forward looking as always, Higginson and his building committee had one more request: that McKim use fireproof construction, a unique concept for concert halls at the time, and which later was shown to be important for the preservation of bass sounds.

McKim came back in about four weeks with drawings that followed these guidelines. Higginson lent the revised plans to Sabine, again without McKim's knowledge, and they met to discuss them about the middle of January 1899. The drawings showed one balcony on each of the two side walls and two balconies at the rear. The width was now 75 feet and, to accommodate the required 2,600 seats, the length was 168 feet, unusually long for a concert hall. Higginson and Sabine agreed that the resulting sound would be like what one would hear in a tunnel. To shorten the hall without sacrificing seats, Sabine recommended the addition of a second balcony on each side. McKim's drawings showed the performance stage at the front of the hall, with the side balconies extending over the left and right sides of the stage. Sabine recommended that the orchestra be placed in a narrower stage house off the front of the hall, thus freeing up space for 200 seats on the main floor. But the hall still seemed too long. Higginson decided to reduce the row-to-row spacing of the seats from McKim's 36 inches to the old hall's 31 inches. He also asked for four rows of seats in the side balconies, one more row than in the Gewandhaus. Sabine, meanwhile, used his formula to calculate optimal ceiling height, critical to determining the desired reverberation time. Higginson sent the modifications to McKim on January 26, apprising him of Sabine's participation for the first time.

McKim was distressed by latest round of changes, which meant that the new hall, far from being an innovative architectural monument, would look very much like the old Music Hall. Higginson arranged for Sabine to go to

New York to mollify McKim: the two men spent about two hours together. McKim ended up accepting the second side balconies and the reduced row-to-row spacing, but, for sight-line reasons, remained intransigent on having three rows of side-balcony seats rather than four. McKim, and officially his firm, agreed to hand over to Sabine all responsibility for the acoustics. He most likely did so for four pressing reasons. First, he suspected that Higginson was implacable; second, there was no time for further argument (only 19 months remained until the inaugural concert); third, he did not want to accept responsibility for acoustics that might turn out badly; and fourth, he wanted to show that the less-than-original interior of the hall had been determined by factors beyond his control. McKim now devoted himself to the interior design, asking for statues (as he had in the original design), coffers in the ceiling as in the old Music Hall, and an abundance of gold leaf and rich red on the balcony fronts and the stage proscenium. For his part, Sabine asked that the ventilation be designed so that air would enter the hall from the ceiling and exit through outlets near the bottom of the side walls. On the basis of his acoustical calculations, he insisted that the ceiling above the stage be lowered—over the strong objections of the organ builder, who argued that it needed to be 10 feet higher to accommodate the full length of the lowest-toned pipes. Boston Symphony Hall opened October 15, 1900, a great success and testament to the fact that every change to the architect's original design made by the building committee, Higginson, the acoustical consultant, and the architect was in the direction of better acoustics.

Design under Way: Philharmonic Hall

The passionate care and conviction that inspired these discussions are integral to understanding how the acoustics of Boston's Symphony Hall ended up, by general acclamation, as among the finest in the world. On the Lincoln Center project, we took some of that same spirit to heart and gained much from what we knew of the techniques and experiences of long ago. We met with Max Abramowitz and his staff, reported our findings, and urged that the hall be shoebox-shaped and not larger than Symphony Hall. Because the row-to-row spacing of the Boston hall (which seats 2,625) would be too narrow for taller patrons, this meant restricting the seating capacity of Philharmonic Hall to no more than 2,400.

By March 1959, Abramowitz presented a design that met our acoustic objectives. Although resembling Symphony Hall, it was wider, 100 feet; it had acoustical reflecting panels over the stage and front part of the ceiling as at Tanglewood. The design showed a seating capacity of 2,400, a volume of 700,000 cubic feet, and a reverberation time with full occupancy of 1.8 seconds at middle frequencies, equal to that in the best concert halls. The stage was 40 feet deep and 61 feet wide, and the motorized overhead canopy could be raised or lowered to suit the musicians' preferences.

On December 2, 1959—under the title "Final Design"—the *New York Times* described our approved design using as illustration the architect's sketch Abramowitz had presented at a luncheon of the Friends of the Philharmonic. Lincoln Center's own house newsletter, *The Performing Arts,* emphasized the need for a hall with no more than 2,400 seats. Had this design carried the day, the hall would have sounded much like Symphony Hall, Boston, and Bolt Beranek & Newman would have been the toast of New York. But several New York newspapers, in particular, the *Herald-Tribune,* campaigned against the seating limit, arguing that the hall should have at least as many as the 2,760 seats in Carnegie Hall, then scheduled for demolition. The Lincoln Center building committee caved in to this pressure and instructed the architect to increase the seating capacity, using whatever means possible. Because Abramowitz wanted no direct contact between BBN and the building committee, I never had an opportunity to debate this change with committee members.

In our original design, the side balconies extended horizontally, parallel to the main floor. In the new design they sloped steeply downward: the front of each side balcony could be entered from one floor and the rear from the floor above. To cram in more seats, these balconies were bowed, a change I reluctantly agreed to, but no one advised me of their steep slopes, then or later. Sometime afterward, my co-consultant Russell Johnson discovered by chance how steep they were when he glanced at drawings sent to Laymon Miller at BBN, who was working on ways to quiet the air-conditioning system. Russell thought we should not worry about those drawings, which were probably prepared for bidding purposes—if they were in play, surely the architects would have sent them to us. Yet after the hall opened, the architect and his staff blamed us because we had not complained about the steepening. Had we expressed our opposition, they said, the change might not have been made.

But for an unfortunate event, I probably could have voiced my opinion about the sloping balconies a month or so later in a meeting in New York, attended by the seating consultant, lighting consultants, the architects, and me. A whole series of items came up for discussion, most of which did not involve acoustics. About a third of the way through the agenda, my feet swelled enormously and, in deep pain, I had to leave. The day before, I had received a large shot of penicillin after cutting my hand, and I turned out to be allergic to it. The sloping balconies may have come up after I left, but no minutes of the meeting were ever sent for me to review.

As if this weren't enough of a problem, yet another cropped up when it came to determining how the panels would be hung over the stage and front of the auditorium. Abramowitz and his staff worried for nearly a year about how to make the panels look good. They never arrived at a satisfactory solution, and Abramowitz finally decreed that the panels be hung over the entire audience area from front to rear. I consented to this with great reluctance, on the proviso that if the arrangement did not work acoustically, some of the panels could be pulled up to the ceiling to simulate the Tanglewood solution. But to my horror, they were all welded together, forming a huge raft, because, at the last minute, Abramowitz worried that they might bang together in an earthquake and come tumbling down.

Still another tough issue involved the design of surfaces on the side walls. For good-quality reverberation, walls and ceiling must have irregularities—achieved in Symphony Hall, Boston, for example, by arrays of niches and statues on the side walls and coffers in the ceiling. Abramowitz intended to include the irregularities we had designed, but the building committee ruled them out to cut costs following a massive overrun. An interior decorator was hired, over the objections of Abramowitz, to cover up their absence, ultimately by painting the walls blue and illuminating them with blue floodlights.

And yet another snag arose at the last minute. The canopy over the stage was to have been hung below the first row of ceiling panels by about six feet, and the front edge of the canopy was to have had a vertical screen to make the design visually acceptable; this meant that the canopy could have been pulled upward six feet above our recommended height if the performers so preferred. But the building contractor misread the drawings and the row of panels nearest the canopy ended up six feet below its intended position. Thus, if the canopy over the stage were pulled upward, a single row of panels would hang six feet below both the canopy and the next row of panels. The orchestra at the first rehearsal wanted the canopy higher, and said so, but when I discovered the error during the week preceding the hall's opening and reported it to Abramowitz, nothing could be done about it at that point. For visual reasons, the motorized canopy had to remain fixed in height—although later, it was raised one foot.

Opening Night Debacle

With all of these deviations, the grand opening of Philharmonic Hall was not what it should have been—a chance for us to beam with pride over a fine accomplishment—but instead a traumatic disappointment, even a nightmare. Music critics came from all over the world and the program was broadcast live by TV and radio nationwide. The performance demanded an extra-large orchestra, along with three large choruses and a host of soloists. To accommodate this assemblage, the stage had to be extended to double its design size. The performance pieces, under Leonard Bernstein's baton, came across as overwhelmingly loud and aggressive, particularly with sound bouncing off the acoustic panels everywhere. One composition called for a rail to be struck by a sledgehammer. The sound was so loud that I went back to the hotel with a splitting headache and the conviction that the reviews would be universally hostile. Performances by the Boston, Philadelphia, and Cleveland orchestras later in the week sounded better, partly because the pieces performed were not so large-scale, partly because the regular stage size was used, but the damage had already been done: most of the music critics had already formed an opinion.

The initial reaction—mixed, cool, somewhat muted in its criticism—was not as bad as I had anticipated, however. Critics in the press complained about a modest lack of bass; some performers on stage discerned a mild echo coming back at them. Maestro George Szell quipped: "One cannot make love in a blue hall!" Maestro Bernstein, on the other hand, dug more deeply. "Everything on stage looks very distant, making it difficult to accurately judge the acoustics. This effect is like looking through the wrong end of a telescope." On the other hand, the Philharmonic's reed players positively glowed about how their sound projected into the auditorium.

Describing the hall in the Sunday *New York Times,* December 9, 1962, music critic Harold Schonberg wrote that it

does have some defects, as the acoustician in charge and the Lincoln Center officials are readily prepared to admit. These defects are of course, under intensive study. The big trouble remains the lack of bass response. . . . Bass aside, the hall has some considerable virtues. Sound is unusually clear: a little dry, perhaps, but certainly not lacking in color. The hall is very kind to solo violin and piano. . . . If the soloist can produce a good sound, it will sound good in Philharmonic Hall. . . . I have grown to like the sound in Philharmonic Hall. . . . When its central problem is overcome, Philharmonic Hall will be one of the world's great auditoriums. . . . [I]t will be a hall of unusual clarity and honesty in which musical strands are never obscured and in which sound has remarkable presence.

Unfortunately, New Yorkers never saw this review. The city's newspaper delivery services went out on strike and the paper was not printed—except for the Arts and Leisure and Magazine sections, which had been sent the preceding Friday to other cities, including Boston, where I found the review.

BBN quickly set about remedying the acoustical defects. Using a model of the suspended panels, we took measurements and discovered that, by simply realigning them and getting rid of the unusual step at the front, we could substantially improve the bass response. The echo could be eliminated by covering the front wall of the projection booth above the second balcony with a better sound-absorbing material. We also concluded that the huge pipe organ behind the orchestra was absorbing bass sound, and that a retractable lead curtain should be hung over it to isolate it from the hall when it was not in use. Seeing no way irregularities large enough to make an appreciable difference could be added to the walls, we made no recommendation in that regard; our recommendations for improving the bass response and eliminating the echo were all approved by the Lincoln Center building committee.

A Vendetta, on the Heels of a Misunderstanding

Enter Maestro George Szell, then music director of the Cleveland Orchestra. A new high school auditorium had been completed a few years earlier in Lakewood, a suburb of Cleveland, and the president of the school board, a friend of Szell's, asked if he would bring the orchestra over to play at the dedication. Szell agreed. Bolt Beranek & Newman got word that he was planning to rehearse a few days before the hall's opening, and one of our men went out to Lakewood. We knew the venue was not good for symphonic music becoming the ceiling was only 15 feet high. Szell didn't like the sound either, so he said to the BBN man: "I think you should hang some absorbing material on the bottom four feet of the rear wall of the stage." If our BBN man should simply have agreed to do this: even though it offered not the faintest prospect for improvement, it would have done no harm. But instead, he was blunt: "I think you'd not hear any difference, Dr. Szell." "Do you think I can't hear?" Szell retorted, snapping his baton in two and storming off.

Szell told the story to everybody, with a twist that showed how seriously he had misunderstood the encounter. The general manager of a hall in the Midwest told of Szell conducting there and condemning BBN for insulting his powers of hearing. Music critic Irving Kolodin came to me in my hotel in New York a day before the opening of Philharmonic Hall, and said: "I talked with George Szell today, and he told me he's going to damn Philharmonic Hall, even though he's never been in it, because you guys said he can't hear."

Around Christmastime 1962, the president of Lincoln Center, composer William Schuman, had lunch with Szell at LaGuardia Airport. Once again relating the Lakewood story, this time Szell excoriated BBN so vehemently that Schuman vetoed the board's decision to go ahead with our recommendations. He instructed Abramowitz to put together an advisory committee to come up with alternatives. The committee consisted of Vern Knudsen from UCLA and Seattle acoustical engineer Paul Veneklasen, who had worked on just one hall each; Manfred Schroeder from Bell Telephone Laboratories, who was to perform acoustical measurements; and Heinrich Keilholz, who was a highly respected *Tonmeister* from a German phonograph recording company.

The advisory committee came up with a recommendation that, far from improving the acoustics, resulted in yet another acoustical fiasco. The hanging panels came down and, in their place, a steeply sloped canopy was positioned over the stage. This panel reflected the sound of the orchestra toward the balconies at the rear; the balconies, in turn, returned an echo that made music in the first dozen or so rows of seats sound terrible. When BBN was shown the recommendation before construction, we warned that this was the wrong direction to go in a letter signed by Richard Bolt, Robert Newman, Ira Dyer, Theodore Schultz, and me, to make clear that our opinion was jointly held.

In August 1963, the editors of *Audio* magazine expressed their displeasure with how things were proceeding: "Unfortunately, from the first performance in this Hall, a hue and cry has been raised which finally culminated in a committee of experts' reviewing the acoustical 'condition' of the Hall and recommending changes. Frankly, we think the procedure in this area has been degrading and unethical."

For BBN, the worst part of the whole episode was the interview given by an anonymous member of the advisory committee to Harold Schonberg of the *New York Times*. This member revealed that he had only joined the committee to "save acoustics" from the miserable failure caused by my incompetence. Tough words indeed, but also ironic in light of the committee's role in making matters worse.

The spring meeting of the Acoustical Society of America took place soon after the appearance of Schonberg's article. I got wind that a move was afoot to rescind my membership. The talk was alarming enough that ASA's president, Paul Boner, called in Knudsen and Veneklasen, themselves ASA members, for a meeting with the society's Executive Council. The two were asked to put a lid on negative chatter, and, if possible, to make a show of professional unity. Rumors also started circulating in Europe that I was about to be fired as BBN president, which did not happen.

Although there was probably enough blame to go around, the troubles at Lincoln Center were clearly not all the result of my mistakes. The hall, I am convinced, would have been improved and at far less cost had BBN's recommendations after the opening been followed. After two more failed attempts by members of the committee to better the acoustics, Philharmonic Hall underwent a major rehabilitation in 1976 and was renamed "Avery Fisher Hall," in honor of its benefactor.

In succeeding years, the Acoustical Society of America has given me its complete support, just as it did when that small clique of members set out to discredit me in 1963. I received the Society's Gold Medal Award in 1975 and, in 1994, became one of four living members with the title of Honorary Fellow. I was asked to serve as cochairman of the seventy-fifth anniversary celebration in 2004, an honor I gladly accepted. After the celebration, the society's thirteen technical sections jointly sponsored a special ninetieth-birthday session for me. Most unexpected of all, the year before, I had been called to the White House to be presented by President George W. Bush with the National Medal of Science. And what an occasion. In the East Room about 200 had assembled: President's cabinet members, government agency members, families and friends of the honorees, past recipients of the award, and a huge media contingent. The eight laureates included two biologists, a chemist, an engineer (me), a mathematician, and three physicists. When Bush's name was announced all stood up as he entered. He gave a speech on the importance of science to the nation and then, after a Marine officer read the encomium, he hung the heavy Science Medal attached to a wide ribbon on each. Afterward, we went to the formal dining room and were served with a lunch while a Marine string quartet provided background music. I talked briefly with Bush and the Secretaries of State and Energy. That evening, a larger group was assembled at the Ritz-Carlton Hotel for a black-tie dinner. After the meal, a video was shown that presented, briefly, the accomplishments of each of the awardees. The medalists are selected by a Committee of fourteen scientists and engineers appointed by the President. The medal is administered by The National Science & Technology Medals Foundation, established by Congress in 1959. Starting in 1962, more than 400 have been recipients.

A Fresh Start

Although other interests kept me away from work on concert halls and opera houses for more than two decades after the Lincoln Center fiasco, in 1986, I decided to return to the field. This was no easy task: acoustical research and technology had moved far ahead of where I had left off. So I committed three years, nearly full-time, to catching up. I went to the library at MIT almost every day; I immersed myself in the acoustical journals that had accumulated in my own library since my active period. I attended concerts—partly for pleasure, but mostly to analyze acoustical effects—in 23 halls in 19 cities: Baltimore, Boston, Chicago, Costa Mesa (California), Dallas, New York, Lenox, San Francisco, London, Cardiff, Glasgow, Amsterdam, Berlin, Munich, Leipzig, Vienna, Budapest, and Christchurch and Wellington (New Zealand). Hearing of my quest, the editor of *JASA* invited me to prepare a review and tutorial paper, which ran 39 pages and was published under the title "Concert Hall Acoustics—1992." After evaluating the acoustics in ten representative concert halls, the paper presented a summary of recent published research on concert and opera halls in Germany, England, Japan, Denmark, and the United States; it ended by proposing four architectural choices (rectangular, fan-shaped, "vineyard," and "arena") and seven essential attributes of good concert hall acoustics (reverberance, loudness, spaciousness, clarity, intimacy, warmth, and hearing on stage). The paper was well received. Little did I know how valuable these three years of study would be until, in 1989, I took on an assignment as Acoustical Design Consultant for the Western-style opera house in the New National Theatre in Tokyo and the Tokyo Opera City Concert Hall—a story told in chapter 10 of this memoir.

Behind the monotony of legal language is an exciting story of intrigue, surprise, big names, political influence, vast power, struggles for survival among huge corporations. . . . Among the directors of the eight major companies which have sought the license have been doctors, judges, scientists, educators, editors, brokers, athletes, engineers, businessmen, lawyers, politicians and enough Choates, Cabots, Taylors, Clarks, Halls, Winslows, Browns and Hendersons to start a new Boston Social Register. As one attorney put it, "anybody in Boston who was anybody was invited to dance."
—*Boston Globe* January 23, 1972

I took part in this dance, risking at one point all of my life savings. The stakes were high—financial independence or bankruptcy—yet even as I seemed to be losing like some compulsive gambler, I kept betting more chips. So did thirty of my colleagues. The story covers nearly a decade of legal dueling: four times to the Federal Communications Commission, four times to the U.S. Court of Appeals for the District of Columbia, and three times to the U.S. Supreme Court. There was much high drama along the way—personal conflicts, twists of fate and fortune, and the inevitable chance discoveries that, if not recognized and acted on promptly, could have sunk the whole endeavor.

Emergent Strategies

It all started in October 1962 over lunch with Jordan Baruch, a partner at Bolt Beranek & Newman. Out of the blue, Jordan told me what had just happened to WHDH-TV Channel 5 in Boston, then owned by a leading local newspaper, the *Boston Herald Traveler*. A few weeks earlier, having discovered that the president of the Herald Traveler Corporation had made ex parte contacts with the commission's chairman during the original hearings which gave WHDH a construction permit, the FCC had awarded the newspaper a license to operate WHDH for just four months, instead of the customary three years. It seemed likely, that, upon renewal, the license would be awarded to someone other than WHDH-TV, and Jordan thought the chances were reasonably good that any like-minded group of committed members could be that someone. Indeed, Jordan had already been invited to join a prospective stockholder group, whose members included several attorneys from the Boston law firm Brown, Rudnick, Freed & Gesmer.

So it was that I found myself in the coffee shop of the Parker House Hotel in Boston with one of those attorneys, Nathan David, on the afternoon of November 13. David briefed me on the WHDH Channel 5 case. President Robert B. Choate of the Herald Traveler Corporation, the parent organization, had taken FCC Chairman George C. McConnaughey to lunch on two occasions in an effort to influence the commission's vote. It was also alleged that a bribe had been offered, as had actually occurred with former Commissioner Richard Mack in the granting of licenses to two Florida TV stations, though the WHDH-TV allegation was never proven. David showed me new FCC Chairman Newton Minow's strong dissent to the commission's majority decision to grant a four-month license; he had voted to deny the license because of the ex parte contacts. When, however, the Herald Traveler Corporation filed for renewal of license, the commission issued a striking statement: "New applicants are invited . . . to file applications in competition with WHDH-TV in a comparative hearing."

After David outlined a strategy to secure the license, I told him I would join this enterprise only if the entire group of potential stockholders were reputable members of the community committed to improving television. "If their only interest is financial gain, count me out." David showed me that revenues at each Boston TV station were more than enough to fund better-quality local programming and still yield an acceptable profit.

A week later, I met with Judge Matthew Brown, who shared my views on what the makeup, temperament, and outlook of the stockholders should be. In mid-December, I called to say I would join the group and assist in the preparation for filing with the FCC. We called ourselves "Boston Broadcasters Incorporated" (BBI).

Before leaving with my family on our annual month-long skiing vacation in Switzerland, I assured BBI that on my return in February I would get right to work on technical and legal matters, and on recruiting others to join our effort. The board asked if I would consider serving as BBI president. I agreed, on the proviso that the job take up no more than 20 percent of my time; I felt that Bolt Beranek & Newman needed at least 80 percent. On my return, I found that I'd been voted president and CEO; Matthew Brown chairman of the board; Nathan David executive vice president; William Poorvu treasurer; and Joshua Guberman secretary.

And what a group we put together. Oscar Handlin was a leading scholar in American history; Robert Gardner an eminent anthropologist and film-maker; Stanton Deland an attorney and the chairman of Harvard's board of overseers; William Andres an attorney and the chairman of Dartmouth's board of trustees; Gerald Holton a professor of physics at Harvard; Henry Jaffe a producer of television network shows; John Knowles the head of Massachu-setts General Hospital; Charles Maliotis, Alfred Morse, and Charles Marran presidents of large Massachusetts companies; Jordan Baruch an electrical engineer; Matthew Brown, Alford Rudnick, and Nathan David attorneys; Edward Bursk a publisher; Constantine Pertzoff a Boston architect; and Wil-liam Poorvu a real estate developer.

Our goal was to create a different kind of TV station, relevant, as our pro-spectus stated, to "the needs, problems and tastes of the community: One that brings the best of Boston's unique local resources as well as those of the nation to television audiences, not by a scattering of programs specifically labeled 'educational,' but by infusing the whole schedule with more exciting and meaningful material." We visualized more in-depth and longer coverage of news and public affairs, free air time for religious and political broadcasts, more local programming than any other station in the United States, and regular editorials based on thorough research and reflection. Above all, we wanted to voice an informed opinion on issues that mattered to the Greater Boston and eastern Massachusetts communities.

There were three other applicants for the license: WHDH (the FCC allowed them to apply for renewal, and during the hearings they acted as though they had never operated a TV station); Charles River Civic Television, Inc., a group of old Boston Brahmins headed by Thomas Cabot of *the* Boston Cabots; and Greater Boston TV II, Inc., a group of Irish citizens with J. Joseph Mahoney as attorney. Our Washington attorneys, skilled in FCC matters, were Fly, Shuebruk, Bloom & Gaguine. Benito (Benny) Gaguine was our lead attor-ney, a short, somewhat stout man, sparse on hair, about 50 years old, with an engaging personality. He liked good food and wine and played poker at least once a week, often with a group that included a U.S. Supreme Court judge. His assistant, Donald Ward, was a young lawyer of about 40, smart as a whip and the best writer I have ever known.

Even though I was swamped at the time with ever-expanding business at Bolt Beranek & Newman, I managed to look into the broadcast equipment we would need for the station. I contacted RCA and acquired from them a complete set of specifications as well as a price quotation, which became part of our filing for an FCC license. I also negotiated a contract with the Westinghouse television station in Boston, WBZ, which would allow us to place our antenna on their tower below theirs. Some might have thought we were getting a bit ahead of ourselves, but we wanted to make a thorough presentation and, at a series of meetings attended by half of the stockholders prior to filing for license, our discussion focused almost entirely on programming content—and who among the stockholders would take part in particular kinds of programs. We were all very gung ho. I agreed to help with science programming and editorials. Our search for a general manager began with consulting David Ives, head of the local PBS station, WGBH, who made a first-rate suggestion. Richard S. Burdick, had distinguished himself as general manager of two PBS stations—WUHY in Philadelphia and WHYY in Wilmington, Delaware. Ives knew that Burdick would like to live in Boston. We chose Larry Pickard, who had been managing editor of NBC-TV's *Today* show and was currently news director at WBZ in Boston, to direct our news department.

License Application and the First Round of Hearings

Our oversized application packet, filed on March 26, 1963, included information on our stockholders, officers, finances, planned programs, engineering data, antenna, and studio site. The next step was for the FCC to appoint an "Examiner," who would hold oral comparative hearings in Washington, D.C. The hearings did not begin until July of the following year. Our examiner was Judge Herbert Sharfman; the opposing counsels were William Dempsey for WHDH, Harry Plotkin for Charles River, and J. Joseph Maloney Jr. for Greater Boston II. I had anticipated that the examiner would want factual information and that, as the first witness, I would be able to refer to notes from the witness box. No such thing. So the hearing turned into a memory test more than anything. Our attorney, Benny Gaguine, took me through my paces in outlining the reasons I had become part of Boston Broadcasters, what I expected to do, and basic goals for the station.

Then came a grilling from the other sides. William Dempsey was the more difficult and aggressive of the counsels. His questions focused on several aspects of television; he especially wanted to trip me up on my knowledge of broadcasting. I answered many of his questions well, but on others my lack of television experience showed. Dempsey wanted to know how I expected to be an effective president of BBI if I devoted just 20 percent of my time to the job. Though it may not have satisfied everyone, my answer was candid and direct: we would have a general manager to run the station; our executive vice president, Nathan David, would be at the station full-time; and I would be called on primarily to settle disputes and to sit on the editorial board. For his part, Joseph Maloney stayed away from questions he knew I could answer and succeeded in embarrassing me by asking me to name the members of the Boston City Council, the Winchester Board of Selectmen, the Boston School Board, and the Winchester school superintendent. I knew a few of these, but not nearly enough to escape Maloney's withering scorn. Harry Plotkin wondered about the duties and responsibilities of BBI's chairman of the board, president, executive vice president, and general manager. He asked sarcastically whether we had discussed proliferating executive officers to make a better impression on the FCC and using a videotape or a live band for the national anthem at the start or conclusion of each broadcast day. I answered yes to the first question, and no to the second.

In my passable testimony, I emphasized that we were a group of citizens who had joined together to bring better television to Greater Boston. Nearly all of our stockholders appeared on the witness stand in the following year, and the other applicants also produced many witnesses.

The Decision: Bad News

The hearings dragged on for more than a year, until July 16, 1965. It then took Judge Sharfman more than another year to issue his decision. When that came down, on August 15, 1966, it was bad news for us. Sharfman ruled in favor of WHDH, the Herald Traveler Corporation. He accused us of overreaching:

It is apt to say about BBI that it went to the extent of impressing the Commission in the comparison by, a) an exaggerated integration proposal; b) an extravagant local live proposal; and c) its impracticable proposed 24 hr operation. . . . BBI . . . lies open to the charge that its "art is too precise in every part." [Its] claims are so nearly perfect in so many areas, that [they] are consistently on the edge of credibility. . . . Admirable, even star-studded as BBI's locally-owned and civically active group is, its promises are permeated by an exuberance which makes one doubtful of their fulfillment.

The judgment hit us like a pail of cold water.

Legal Appeal

Boston Broadcasters called a stockholders meeting to discuss whether we should appeal the decision, who the commissioners were who would rule on our appeal, and what our prospects were for prevailing in the end. Two of the seven commissioners had recused themselves because of prior involvement in WHDH matters. Of the remaining five, Chairman Russell Hyde was a conservative Republican Mormon who tended to go along with the majority opinion. Benny Gaguine was pessimistic about Robert Lee, an accountant, and also a conservative Republican, although the ex parte matter might sway him, and believed James Wadsworth to be totally unpredictable. On the other hand, Benny was optimistic about Robert Bartley, a Democrat who had served many years on the commission. Then there was the maverick, Nicholas Johnson, a 32-year-old, rough-hewn Swede from Iowa, feared by the broadcast industry. Because Johnson usually favored change, however, he represented a possible vote for BBI.

The cards, therefore, were not exactly stacked against us, although most of the stockholders believed there was little chance for a reversal. But the cost of going ahead was only about a tenth of what we had already spent, so why not take the chance? Nathan David estimated the probability of success at about one in ten. When I piped up—"Well, it's still better odds than the Irish sweepstakes, and the prize is much larger!"—everyone laughed, and the group unanimously voted to authorize Benny to file "Exceptions and Request for Oral Argument" with the FCC. The other three applicants also filed.

Because such petitions were automatically granted, all four applicants appeared before the five commissioners mentioned above on September 6, 1967. Each side was given 30 minutes to present oral arguments. Benny waxed eloquent: "This is a group that doesn't need every nickel they can get. . . . If you look at them, you can see that they are the kind of people you can rely on."

We Win!

The big decision arrived a year and half later, on January 22, 1969. The seven members of the FCC were again reduced to the same five mentioned above. Surprise of all surprises—Bartley, Wadsworth, and Johnson voted for BBI; only Lee voted against, and Hyde abstained. The majority concluded: "We view [Judge Sharfman's] findings as warranting substantially different conclusions and a different ultimate result. . . . our judgment is that grant of the application of Boston Broadcasters, Inc., would best serve the public interest, convenience and necessity."

This was quite a turn of events. On January 26, the *Boston Sunday Globe* carried a "Profile in the News." Under the headline "Boston Broadcasters Beranek Appalled by Quality of U.S. TV," the writer described me as an international authority on acoustics, president of Bolt Beranek & Newman with annual income of $10 million, and a mover and shaker in Boston's cultural circles. I had been interviewed for this piece just before leaving for Switzerland with my family. "I believe," the profile quoted me as saying, "that much of the feeling of dissatisfaction in today's young people is caused by so much going on in the world that does not seem related to national, local and societal needs that they protest. We would like to establish that relationship in Boston through our programming." The profile writer added: "At 54, [Beranek] has a knack for communicating. A colleague said yesterday that 'Leo is perhaps the greatest motivator of young men that I know.' That is one reason why Beranek has been able to assemble such a bright group of young men around him at BBN, the company he helped found 20 years ago." A generous compliment, to be sure, and one that I had spent my life trying to live up to not only at BBN but in all my ventures.

A New Relationship with Bolt Beranek & Newman

On Monday night, January 27, just after our arrival in Saint Moritz, Switzerland, a call came in from Sam Labate, BBN's executive vice president; Dick Bolt, our chairman of the board; and John Stratton, our treasurer. The tone was concerned, even ominous. John said the Boston Broadcasters publicity—at least my involvement in it—was bad for BBN, and he wondered how it might affect BBN's future. Sam spoke of a "feeling of unrest" among BBN staff.

He said he had contacted Nathan David to ask whether there was any chance that BBI would let me resign its presidency. No way, David had said. Such a move could suggest a falling apart of the organization, which might in turn have an adverse impact on any decision a Court of Appeals might hand down. Sam suggested that I resign the BBN presidency instead, and become a staff member.

I was shocked at this intrusion on my vacation, and it puzzled me that Dick Bolt did not enter into the discussion. I wondered if he wanted the presidency. But there was merit in their position. Plus, I had been president for sixteen years, and a change might benefit BBN for other reasons. The change might benefit me, for that matter, and a full-time commitment to Boston Broadcasters might be just the thing. So I resigned a month later and became BBN's chief scientist. Sam Labate replaced me as president. For the next half year, I was full-time staff on BBN's payroll, then from December 1, 1969, to June 30, 1971, I was carried half-time each on BBN's and BBI's payrolls. On July 1, 1971, I resigned altogether from BBN and joined Boston Broadcasters full-time.

Developing a Solid Base for Boston Broadcasters

In the months ahead, money was a constant issue for BBI. We needed and received, in a remarkable show of confidence in our mission, millions of dollars from our stockholders in cash and in pledges of securities and bank holdings at the First National Bank of Boston as collateral for loans. Equally critical was establishing the framework for an effective TV station, along the unorthodox—some might say "radical"—lines we envisioned.

I set down several objectives for myself. First, I had to be sure that all of us were intent on a schedule that would get us on the air at the earliest possible date because, once that happened, the chances of the Herald Traveler getting us off the air would be minimal. Joining with me wholeheartedly in achieving this goal was William Poorvu, our treasurer, the only trustee other than Nathan David devoting a large amount of time to the endeavor. Second, I had to take an active role in hiring staff, to be drawn, I hoped, from among the best in the country, to operate the station after we went on the air. Third, I wanted the station to be an employee-friendly workplace, with ample health insurance, life insurance, and retirement benefits. Fourth, I believed, along with Poorvu, that we should borrow as much as possible without diluting our stockholders equity; by the same token, we needed to be thrifty in incurring debts, in extending payment dates as far out as possible, and in avoiding interest on delayed payments as long as possible. Fifth, in addition to observing the rules and regulations of federal and state oversight bodies in all stock matters, we had to ensure that we in no way misinformed or under-informed our stockholders about BBI's condition or legal status.

All this involved a lot of maneuvering—and with relatively little wiggle room—in our atypical commercial enterprise, but I felt more than up to the task and even invigorated by it. Between July 1970 and March 1971, I solicited over $300,000 from the stockholders. After long negotiations by Bill Poorvu and me, the First National Bank of Boston agreed to extend us $4 million in credit. Of this, $250,000 would be made available as soon as the U.S. Supreme Court had denied certiorari—that is, refused to take the WHDH case—provided the loan was guaranteed by the stockholders, but no collateral would be required. The next $500,000 would be made available as soon as needed, with stockholder guarantees—but only if the bank saw no substantial reason for doubting we would receive permission to go on the air.

In its January 22, 1969, decision, the FCC had neither actually issued permission for going on the air nor given us a date when it would do so. Further, Benny Gaguine felt that the January 22 decision, which made no mention of the ex parte contacts by the president of WHDH, would not withstand review by the U.S. Court of Appeals. So he petitioned the FCC to supplement and clarify its decision. Observers called Boston Broadcasters "a sore winner." After this request, the FCC added "Paragraph 40" to their decision, which stated in part: "Not until September 1962 [after operating for 12 years] did WHDH receive a license to operate its television station, and even then its license was issued for a period of four months only because of the Commission's concern with the . . . inroads made by WHDH upon the rules governing fair and orderly adjudication."

The Ever-Present, Plodding Legal System

As expected, the Herald Traveler Corporation appealed to the U.S. Court of Appeals in Washington and on November 13, 1970, the court upheld the FCC citing "no error." Incensed, Harold Clancy, president of WHDH, declared: "I think we have been the victim of the most unconscionable injustice in the history of the FCC." WHDH filed for reconsideration, which the Appeals Court denied but granted a window of 60 days in which they could file an appeal with the U.S. Supreme Court. And WHDH did so, arguing primarily that the FCC vote was not legal because only five of seven members had appeared at the final deliberation and only three of those had voted for Boston Broadcasters; three votes, they contended, did not a majority make.

By this time, the proceedings were starting to remind us of Dickens's satires on the worst meanderings of the English legal system during Victorian times, but we kept moving ahead. After considerable effort, we at last found a suitable building for our studios. At a board meeting on March 9, 1971, Bill Poorvu announced a building located at 16 and 37 Permil Road, on Route 128, owned by the Caterpillar Machinery Company near Muzi Motors in Needham, would soon be available. With a ceiling height of two stories in a third of the structure, it could readily be converted to house our broadcasting studios. We immediately bought the building, for $1,050,000. (I later arranged to have Permil Road renamed "TV Place," and our new address became 5 TV Place.)

But we found ourselves in an impossible situation. We did not yet have a valid construction permit, and FCC regulations stated that no part of a broadcast facility could be built without one. Through our Washington lawyers, we requested permission to begin broadcasting in mid-September, the start of the networks' fall season. But this was only six months away, and we had neither hired an architect nor decided what broadcast equipment to buy. However, even without a permit, we were allowed to tear out parts of the building's interior to make it ready for renovation, and to lay the foundation for the transmitter building. It fell to Bill Poorvu to hire an architect and plan the needed renovations. As for the equipment, we needed an engineer with recent TV broadcasting experience. I persuaded WBZ-TV's William Hauser, highly regarded in the industry and newly retired, to extend his working life a few months. When I learned that David Allen, the expert at RCA who had helped us make a list of equipment for our filing, had left RCA, I hired him as well.

Technical Hurdles

Little could match, for sheer drama, the series of events leading to our purchase of nearly $2 million in electronic equipment. To learn what was available and to lay the basis for the best deal, we invited the leading suppliers of television equipment to Boston, first separately to gauge our needs, and then together, on March 18, and 19, to explain why we should select their equipment. General Electric, Norelco, RCA, Gates, and Ampex made impressive presentations. Afterward, we met again with them individually to discuss our needs in light of what we had learned. A week later, Dave Allen and I flew out to the Annual Convention of the National Association of Broadcasters, in Las Vegas, to study the various exhibits of TV equipment. Each of the same five suppliers gave us two hours of demonstrations on their latest lines. By the end of the convention, we were confident in our knowledge of equipment models and types, what they provided, and how to operate them. We were now ready to talk the specifics of design and to negotiate a price.

Bidders' day was April 9, 1971. Dave Allen, Bill Hauser, and I opened the meetings at 9:30 AM in the Hyatt Hotel at Charles River Plaza in Boston. The same five companies made their presentations, again each taking about two hours. Norelco (in combination with Ampex) and RCA made the two best offers. The RCA offer came in at lower cost, but Norelco/Ampex's equipment was of later design. Near the end of the afternoon, we called our managers, Robert Bennett and Richard Burdick, and gave them the details. Both said they preferred the Norelco/Ampex products. But the prices were so far apart we felt compelled to go with RCA. Although my colleagues thought we should tell the RCA people immediately they had won the bid and go home, I suggested we reflect more carefully over dinner.

The three of us sat at a table in the corner of the hotel restaurant while the bidders sat at their separate tables, casting nervous glances our way. Near the end of the meal, Dave excused himself. Returning in a half hour, he informed us he had just talked in the hall with the Norelco/Ampex representative and had hinted to him that they were possible losers. The representative said that they would seriously consider reducing the bid, if we could give them a couple more hours to communicate with their headquarters. We let the other four companies know we needed more time for study. In the end, Norelco/Ampex came up with the lowest price, $1,775,000, and we signed a contract with them on April 27.

Meanwhile, the wheels of justice labored on in the background. We rejoiced when on June 14, 1971, the U.S. Supreme Court declined to take the case, denying WHDH's petitions for writs of certiorari. But this was just the first of three times that the Court would be asked to consider the case over the course of the next nine months. We pressed ahead regardless.

Spreading Myself a Bit Thin

The year 1970, working half-time at Bolt Beranek & Newman and half-time at Boston Broadcasters, had been exceptionally busy. At BBN, we brought ARPANET—the first leg of today's Internet—into operation. Visitors from universities who wanted in on the network, programmers seeking employment, venture capitalists, and foreigners interested in what we were doing began pouring in and had to be welcomed and entertained. Because of all the computer-related research contracts we were acquiring, we were constantly recruiting and hiring. Spending my spare time for months beforehand copying examples to tape and assembling motion picture material, I spoke in February to an audience at the Swiss Federal Institute of Technology on the history of the use of computers to generate speech and music; the lecture was well attended, and a transcript appeared that December in the *IEEE Transactions on Audio and Electroacoustics.* I also gave a second lecture, also well attended, at the Swiss Federal Institute, on concert hall acoustics, the next day and again in Paris in April.

Back home, at the 1970 Spring Meeting of the Acoustical Society of America, held in Atlantic City, I gave a speech in honor of Ted Hunt, my old Harvard professor, who had been selected for the society's Gold Medal Award. A week later, I flew to Los Angeles to preside, as president, at the meeting of the Audio Engineering Society. Then it was off to my alma mater, Cornell College in Iowa, for the spring meeting of the board of trustees and for a milestone in our lives—our son Jamie's graduation from Cornell.

Soon afterward, six of us at Bolt Beranek & Newman went to Washington, D.C., to make a presentation on computer technology for the air traffic system to the U.S. Department of Transportation, which, unfortunately, did not lead to a contract. The week after that, we had had a key and this time successful meeting about leasing our time-shared computer services to Harvard University. In mid-June, I flew to Dallas for two days of meetings with my old boss, Arthur Collins, president and CEO of Collins Radio to pitch our computer services, unsuccessfully as it turned out. Then there were meetings and discussions with Boston Broadcasters board members and officers whenever I happened to be in Cambridge and with fellow council members of the National Academy of Engineering in Washington, where I also served on three committees.

In the midst of all this, in early July 1970, I was asked by President Nixon to chair the Supersonic Aircraft Neighborhood Noise Committee for the Department of Transportation. The newspapers had been filled with controversy about the noise a supersonic jet plane might make. After traveling to Seattle on several observational and data-gathering visits, we had to coordinate our findings with those of other committee members working in other areas. In the end, we were able to get Boeing Aircraft to change its type of engines and to add mufflers. I spent an enormous amount of time on this project, preparing my testimony about the expected noise from the supersonic transport (SST) for hearings before the relevant House and Senate committees that fall and on into the following spring. One week, I went to Montreal on BBN business, and the next to Chicago to testify before the City Council on a proposed noise ordinance. In the meantime, I took part in regular policy discussions of the Massachusetts Commission on Ocean Boundaries. In early October, I flew to Saint Paul to attend the opening of a new concert hall designed by BBN, with the Minneapolis Symphony performing. On nights and weekends, I worked away at what would be one of my most successful books, *Noise and Vibration Control,* completing the manuscript in mid-1971, just as I went full-time at Boston Broadcasters.

All through this exceptionally busy period, I tried to spend what time I could with my family. We had tickets for a series of Theater Guild plays and season tickets for the Boston Symphony Orchestra. We attended concerts and other events in Winchester, and sometimes went to meetings of the selectmen. Phyllis and I had a standing date for dinner in a leading Boston restaurant once a week, becoming regulars in at least three of Boston's better-known restaurants.

Focus on Boston Broadcasters

Bill Poorvu and I were beginning to see that Richard Burdick was not going to make a good general manager for BBI: he lacked experience in handling a sales department and in purchasing syndicated material. Changing Burdick's title to "Manager of Creative Services" freed us to hire a general manager for operations who had experience in a running a commercial station. Bill and I first interviewed one candidate, Robert Bennett, then general manager of Channel 5 in New York, who had also run Channel 5 in

Washington, D.C., on October 6, 1970. We found him exceedingly bright, eminently qualified, and almost certain to learn quickly in any new situation. BBI's board of directors met on January 14, 1971, to make a selection from among four candidates. We interviewed three one day and Bob Bennett the next morning. Bob came on strong. He clearly knew how to manage a station. That he had raised the news programming on Channel 5 in Washington from number 3 to number 1 in the ratings and had run a first-class operation in New York convinced us that management and sales were his strengths.

Now came decision time. A split on the board emerged. Half wanted Bennett and half wanted one of the other three, Fred Walker, who was strong on programming. During a brief break in our deliberations, Bill Andres took me aside and asked which man I really wanted. There was no doubt in my mind, I said, that we needed a man with Bennett's experience. When we reassembled, Andres argued eloquently that the board ought to give the president whatever he felt was necessary to get the job done. As a result, all but one vote out of eight went for Bennett.

The Battle Grows Ugly and Very Public

Meanwhile, acutely aware of how far along Boston Broadcasters was in finally getting on the air, WHDH went on the legal offensive once again—this time with a vengeance. They assigned between nine and thirty of their newspaper reporters to investigate every aspect of our history and shareholders, with special emphasis on Nathan David, our executive vice president. Then they filed a petition with the FCC alleging, as reported in the April 26 issue of *Broadcasting* magazine, "that Mr. Nathan David was guilty of misrepresentation, fraud, and violation of Massachusetts and Federal Security laws in connection with the sale of 6,000 shares of unregistered stock in Synergistics, Inc., a firm with financial interests in CATV." WHDH also accused us of having willfully withheld from the FCC our decision to hire Bob Bennett as general manager for operations and our purchase of more than $3 million worth of TV station equipment. In point of fact, we were in the process of filing the required information on both matters—well within the FCC's deadline—but the charge against Nathan David was serious.

I lost little time in putting together our own press release: "WHDH's latest petition is a desperate and last ditch effort to save a sinking ship. . . .

We will answer each and every charge at the appropriate forum before the FCC." WHDH's president, Harold Clancy, shot back: "I am pleased that Mr. Beranek—who did not deny a single charge in his press statement—plans to make his answer to the Commission. This is precisely what we have asked the FCC to do; to investigate our charges and determine their truthfulness in an appropriate Commission proceeding." The last thing we wanted at this point was to get caught up in legal wranglings, so we relied on our trusted counsel, Benny Gaguine, to file our response, which he did on May 11. We pressed on, irritated but undeterred.

The exchange between me and Clancy was the first of four occasions when our statements appeared, in columns side by side, on or near the front page of the *Boston Herald Traveler,* which catered to the business community and the Boston Brahmin set. Exposed to Clancy's accusations and the row raging in their morning papers, some of my friends and colleagues wondered how this could possibly be the same Leo Beranek they knew.

We Are Granted a Construction Permit

On May 19 we petitioned the FCC for an on-air date, asking for 3 AM one day in mid-September, if the Supreme Court denied WHDH's request for a hearing. Our most urgent need at this point was a construction permit, which the FCC finally granted on July 23, just two months before we planned to go on the air. I issued a terse, public statement, carried by the media: "Boston Broadcasters, Inc., now has in its possession a construction permit issued by the FCC . . . authorizing BBI to construct a television station to operate on Channel 5 in Boston. The FCC also issued the projected station's call letters, WCVB-TV. BBI is now free to turn its attention to the finalizing of the many details essential to putting Station WCVB-TV on the air." Harold Clancy's reaction was predictable. "Mr. Beranek's statement . . . appears to me to be another effort to confuse and mislead," he fumed. "As Mr. Beranek well knows, this action was merely a ministerial act which in no way alters or changes the Channel 5 situation. Mr. Beranek is free to build at his own peril." As before, his and my statements appeared in parallel columns in the *Herald Traveler*—intended, no doubt, to hurt our case in the eyes of both our bankers and TV viewers in Boston.

With the construction permit in hand, we moved ahead quickly. On Bill Poorvu's urging, we established the Hawthorne Construction Company and hired Dan Prigmore as its general manager to contract for and to supervise building construction and to install the electronic equipment. In a matter of days, our Needham site was swarming with electricians, builders, and delivery trucks. We asked Bennett, Burdick, and Pickard to report for work immediately, and, when they arrived, to start hiring station personnel to go on the air about September 12.

I remained occupied with contracts, hiring consultations and interviews, arranging deliveries of equipment, and helping where necessary with problems surrounding the renovation of the building and the installation of the antenna on the transmitter tower. On August 12, I advised all stockholders in writing that we must take down the second stage of our loan from First National Bank, amounting to $500,000, and that each stockholder must guarantee the loan to the extent of his pro rata share.

Network, Staff, and Union Issues

When we spoke with executives at CBS Television about carrying their network programs on our station, they expressed their concern that, because we envisioned so much in the way of local programming, many of their programs would be preempted. Indeed, far from considering our proposal seriously, they told us that CBS might very well transfer its network programming to Channel 7 in the event we got the license. Bob Bennett, Benny Gaguine, and I met with two CBS executives in New York, but to no avail. We first felt confident that if CBS went to Channel 7, then ABC would automatically come to us. But ABC expressed the same concern as CBS, that we would preempt too much of their programming. They spoke of putting their network, instead, on three Boston and Worchester UHF stations. In what turned into a frosty discussion, Bob Bennett and I met ABC executives in New York—they seemed already to have made up their minds. A week later, Bob called newly-appointed Vice President Elton Rule to convince him that it would be a major mistake on ABC's part to go with the UHF stations. Bob had been very successful with Channel 5 in New York, after all, and he stressed that we would put on the best news program in Boston—and that, because ABC did not begin its programming until 11:30 AM in those days, we wouldn't need to preempt as much time as for CBS. Bob convinced Rule, who in turn persuaded others at ABC. We got ABC's official go-ahead on August 9.

Another tough question was how we would put together a large enough engineering staff in time for "opening day." We wanted to hire WHDH engineers, who were well trained and wouldn't need advance salaries because they could continue with their present jobs up to the final turnover (we wanted to avoid a big payroll before going on the air). We met with union leaders who controlled the WHDH group, and hammered out a tentative agreement under which the engineers would join us the day WHDH went off the air, and could be trained immediately in their off time.

We also had to engage a national sales representative. As was and probably still is customary in the business, we would set up our own sales force for merchandising of commercials to companies within about 50 miles of the station, but for sales to corporations around the country, we would depend on firms with offices in principal cities outside the 50-mile radius. Bill Poorvu and I spent many days in the offices of five national sales representatives, all in New York, and we finally chose Harrington, Righter & Parsons.

Confusion and Conflict

On July 24, 1971, the FCC issued a perplexing statement: "WHDH may continue broadcasting until further notice." Once again, parallel columns by Clancy and me appeared in the *Herald Traveler*. "The simple fact," I wrote, "is that the Herald Traveler lost its case for renewal of its license by the 1969 decision of the FCC. This decision was affirmed by the Supreme Court's refusal to review the case a month ago. . . . WHDH is being allowed by the FCC to continue broadcasting until BBI commences operations." And Clancy's rejoinder: "Our operation will continue until the FCC acts on the charges we have made and not, as Mr. Beranek says, until BBI commences operations on Ch. 5, because I do not think that day will ever come."

On Friday morning, July 30, a heavy shoe dropped. The headline in the *Herald Traveler* read: "5 men charged by SEC in stock sales probe." Our own Nathan David was mentioned as one of the five, a potentially explosive problem for us, as one of the criteria for the FCC granting a broadcast license was the integrity of the owner's stockholders. A week later, *Broadcasting* asked us if we would agree to be interviewed by Rufus Crater, one of their leading reporters. We accepted, and for us the resulting article could not have appeared at a better time, putting on the record an unbiased, comprehensive account of the state of our construction and plans to go on air. The piece proved highly informative, in fact, to the FCC in later deliberations.

But before the article could appear, yet another threat loomed. Minutes before Crater arrived to interview me on August 16, I learned from engineers at Philips Broadcasting (from whom we were purchasing much of our equipment) that *Television Digest* had come out that very morning with an article under the headline, "Boston's BBI Due for Freeze at FCC." The article was read to me over the telephone, and I tape-recorded it:

Commission is expected to put construction permit of Broadcasters, Inc., on ice pending court and SEC hearings and decision on qualifications of BBI executive VP Nathan David in SEC stock sales case. . . . FCC leans to view that SEC charges against David, if sustained in courts, would disqualify BBI. . . . If Commission sets BBI for hearing on David's qualifications, it undoubtedly will tell BBI to quit spending in preparation for imminent start, in the light of the fact that court and SEC proceedings will take a mighty long time—probably years.

Staying Calm with the Roof Falling In?

I did my best to stay calm and focused as I talked with Crater, even with this new bombshell fresh in my mind, but as soon as the interview ended I called Benny Gaguine, who had not yet seen the article. My heart sank, too, when Benny informed me that his right-hand attorney, Don Ward, was away on vacation and that the matter was too complicated for him to resolve expeditiously on his own. Then a half hour later, to my utter surprise and relief, who should walk through the door of my office but Don Ward. Packed into a station wagon, he and his family were returning from a Canadian camping trip when Don had the bright idea of stopping by to show them the station.

When I played him the tape, the normally coolheaded attorney became alarmed. The first thing we decided was that Nathan David should take a leave of absence and declare he would not vote his stock—and he should do both that very day. By now it was late afternoon. Don decided to go back to Washington right away and, family in tow, drove nearly all night to prepare papers on Tuesday for filing the next day the moment the FCC opened its doors for their usual Wednesday morning meeting. Before getting back on the road, however, Don called Nate, who, in conference with us at his lawyer's office that evening, agreed to take a leave of absence and to forgo voting his stock.

The next morning, with Nate's letter in hand, I grabbed a flight to Washington, where I prepared a statement completely describing the studio and transmitter buildings and affirming that all the equipment would be installed no later than August 31. Benny stated in his brief that we expected to file for "Program Test Authority" (start of on-air operations) on September 12, that we had committed $4.5 million to the project, that we had signed an affiliation agreement with ABC, and that we were moving along satisfactorily with union negotiations. Our filing ended on an urgent note: "[A]ny substantial delay in specifying a termination date, and in granting WCVB-TV Program Test Authority could well completely destroy BBI[, causing] tremendous personal losses to the stockholders [and] havoc [in] the lives of more than 45 employees who have left other positions."

That we were much farther along in construction than anyone outside could have guessed almost certainly influenced the ensuing FCC decision, which was not as grave as *Television Digest* had predicted. Basically, the commission asked the U.S. Court of Appeals to send the case back to it for reconsideration and to inform Boston Broadcasters that it could proceed to construct the station only at its own risk. In my column in the August 21 *Herald Traveler,* I declared: "The FCC's action is without precedent and BBI is confident that the Court will not accede to the FCC's request." In his parallel column, Clancy responded with barely disguised glee: "It is perfectly obvious that the FCC is concerned and wants to investigate BBI's qualifications. . . . I think it is good news for the community we have served for so many years." On August 30, we filed a protest with the FCC, arguing that, because the Supreme Court's action had effectively closed the case, Boston Broadcasters should be allowed to go on the air.

Our First Transmission

The next three and a half weeks were a whirl of activity. On August 24, I wrote to the stockholders of Boston Broadcasters explaining that, because our station was so near completion, we would stick with our original September on-air date. "Our legal position," I observed as soothingly as I could, "is such that there should not be an undue delay, and the completion of construction with reduce the likelihood of delay." On August 24 and 25, I went to the West Coast with a group of our engineers on Bolt Beranek & Newman business, to pique interest at McDonnell Douglas Company in using BBN on one of their quiet plane proposals. And on August 26, I met until noon with the BBN board—this in the midst of intense planning for completion of the TV station and installation of our broadcast antenna on the WBZ tower.

In another filing with the Court of Appeals on September 2, Benny Gaguine called the WHDH request "unjustified, unwarranted, and contrary to the public interest." An appended affidavit prepared by me summarized the status of the station, along with financial statements showing BBI's irreversible commitments at nearly $4.5 million.

Then came the big day, September 10, 1971—risk or no risk, we were going on the air, although for only a few minutes in the early morning, after WHDH went off. The broadcast transmitter had been installed and the antenna mounted safely high upon the tower, with a long concentric cable supplying the crucial link. My son Tom and I stayed up at home to witness our first transmission at 3 AM. About midnight, a call came from BBI's transmitter building. "We're having trouble," said the panicked technician on the other end. "The lightbulb has burned out in our projector and we won't be able to put anything on the screen when we go on air—help us!" I called around and after an hour located a proprietor willing to go out to his store at that ungodly hour to sell us a $1 bulb.

A photographer arrived at our house about 2 AM. At about 4:55, our logo "5" flashed across the screens of television sets throughout Boston with the words "WCTB-TV, Boston, Massachusetts." A photo of Tom and me glued to our set made the newspapers. By September 12, we had completed the six hours of on-air tests necessary before we could file for authority to start full broadcasting. The tests all had to be run in the wee hours of the morning, when WHDH was off air. Tom and I stayed up for some of those, too.

It was last-resort time for WHDH, which quickly appealed again to the Supreme Court asking that its decision be reconsidered because of new evidence—the charges against Nathan David—that might make them think about the case in a different way. The U.S. Circuit Court of Appeals in Washington delayed its own response to the FCC in deference to the Supreme Court. On October 12, the Supreme Court spoke for a second time, denying WHDH's petition for reconsideration. A week later, the FCC almost sealed the outcome by affirming to the Court of Appeals that, even though Nathan David's role had likely been a factor in Boston Broadcasters receiving preference for the license, there was no evidence that BBI had compromised the integrity of the administrative or judicial process in any way.

All equipment was on board and the wiring complete when we were faced with yet another hurdle: the suppliers demanded payment—and we had no money. Bill Poorvu and Dan Prigmore made it clear to me that we needed at least a million dollars to keep suppliers from forcing us into bankruptcy. We laid off about a dozen employees to cut our costs, and we reduced all salaries. At this point, maintaining strictest secrecy, we made a private offering of stock to our existing stockholders. Because of the speed required, I took primary responsibility for preparing a legally necessary offering brochure or prospectus. Working night and day to whip it into shape, with help from our lawyers, I completed it within a week. The stockholders met on November 10, and we raised nearly a million dollars from them. We had the cash in hand by early December.

But yet another problem cropped up. I had to notify our stockholders that the First National Bank was asking for collateral to cover the guarantees they had made on the $500,000 loan. This meant that stockholders had to put up privately owned stock, bankbooks, or whatever, which the bank could sell to cover the loan in case we defaulted. There was nothing we could do but comply, even though we had hoped to avoid this all along.

A Happy Ending?

On December 29, we received a belated Christmas present from the Court of Appeals, whose opinion read:

What ultimately convinces us that the interests of justice would not be furthered by recalling the mandate is the finality of the FCC's order and the award of the construction permit. . . . The commission is bound to respect the governance of a final administrative decision. . . . We see no basis for a claim of unconscionable injustice in permitting BBI's retention of the award on the protective provision of effective separation from Mr. David. The Commission's Request for recall of our mandate is denied.

The media descended on us at 5 TV Place. I put together a statement and read it on the evening television news, adding: "This is a significant occasion for us and a wonderful opportunity to wish everyone a Happy New Year." Our employees were near-delirious with joy. The FCC, I was confident, now had no option but to give us permission to start broadcasting. Of course, the *Herald Traveler* carried yet another set of parallel columns. I wrote of Boston Broadcasters, Inc.'s "long-awaited opportunity to bring New England a new and distinctive television service." To which Clancy responded: "The degree of reliance that can be placed on statements emanating from BBI management can be determined speedily from the groundless representations that were circulated by BBI in June, July, August, September and October . . . etc."

My optimism was confirmed on January 21, 1972, with the issuance of a new ruling by the FCC: there was "no choice under the law, but to grant the request of Boston Broadcasters Inc. (BBI) for program test authority [equivalent to a license to broadcast, but for an interim period of two to three years] for station WCVB-TV, Channel 5, Boston Massachusetts. [B]roadcasting can start March 19th at 3 AM."

The media response came almost instantaneously, and was quite spectacular. The *Boston Globe* carried a front-page story, as well as a personal-interest piece prepared by *Globe* staff writer, Christina Robb, who had come out to the station to interview me. Her article included a rather large picture of me smiling, with the caption "L. L. Beranek 'Momentous Occasion,'" and with a headline 3 columns wide: "Yes, March 19, 3 AM, Brings Grins to BBI." "We heard them scream in Dr. Beranek's office," Robb quoted our telephone operator as saying, "and that's how we found out." "Dr. Beranek," the article went on, "was obviously excited and happy as he greeted the children of excited employees who roved the halls of the never-aired station in laughing, singing, family groups." Robb wrote of the "phenomenal sense of optimism" that bolstered the staff to the end of their wait. The lead article supplied further details, observing that the case had dragged on for 17 years, with BBI involved for nearly a decade. This, the writer said, made it the longest administrative case in U.S. history. Running a close second was the "Carter's Little Liver Pills" case, 16 years, which ended with a ruling that the word "Liver" be dropped; a distant third, 13 years, was the "Peanut Butter" case, in which manufacturers and the government argued over the ratio of peanuts to paste required to avoid the label "imitation." An enormous picture of me appeared, 4 columns wide and 6 inches high, and another with Poorvu, Bennett, Burdick, and Pickard conferring around my desk. *Variety* ended its long article with a comment—"The Boston Channel 5 case is about ready to go down in the books as an historical freak"—that widened yet more eyes over the glacial pace of the legal system.

The board of directors of Boston Broadcasters had met on January 26 to discuss finances—we were running out of money. Bill Poorvu and I were authorized to approach our national representative, Harrington, Righter & Parsons, for a loan of $300,000, and to ask the First National Bank of Boston for an increase in our line of credit from $4 million to $5 million. Many station slots remained to be filled. We desperately needed on-air personnel, and it was our considered opinion that the smallest loss of viewing audience would occur if we could keep WHDH's anchor people. But Clancy had told them all that he would get rid of them if they were caught talking to us. He made good on his threat in one case, when an employee dared to meet with Dave Allen and was fired on the spot. But the on-air personnel all belonged to the American Society of Composers, Authors, and Publishers (ASCAP), and we arranged to meet with them and their union leaders clandestinely in automobiles parked off Route 128. We had to offer stock as well as salary increases, but Bennett and I prevailed and the top anchor people at WHDH came to work for us.

A Final Skirmish

Believe it or not, we were not yet home safe on the legal end. On March 4, the more-than-persistent Herald Traveler Corporation took the case back to the U.S. Court of Appeals one more time, urging a stay on our broadcasting rights until such time as new arguments could be heard. The stress of all this was beginning to take its toll on me. Each time something appeared in the news, I would call Benny Gaguine—or he would call me—and we would discuss our next move, anticipate our next line of defense. Benny was always upbeat, a morale booster if ever there was one, fully confident that everything would go our way. I tried to convey this buoyancy to our staff in the morning briefings, without misleading them into believing we were out of the woods.

Then, on February 27, came another surprise. I had an unexpected visit from a friend of mine who happened to have just made a sales call at WHDH at the very moment Abe Fortas, a former justice of the U.S. Supreme Court and then head of a powerful Washington law firm, was leaving Harold Clancy's office. My friend wondered what Fortas could possibly have been doing there. I said I didn't know, thanked him, and ushered him out of my office. I immediately called Benny Gaguine to sound him out on this alarming bit of news. We suspected that Fortas was about to file for a stay on our case until such time as the Supreme Court could review the latest brief filed by WHDH. Benny said that it took only one justice to issue such a stay and Fortas—an insider who had resigned from the Court just three years earlier—still had more than one friend among the justices.

Even the unflappable Gaguine felt edgy about this turn of events. The next day, he hired an expert on Supreme Court matters, who advised us to have a filing ready to rush in at a moment's notice when word came that Fortas had filed for a delay. Benny prepared a brief asking that the full Court, not just one justice, consider whether a stay should be granted. And we were right about Fortas. He filed on March 13, adding yet a new wrinkle to the case—that the Herald Traveler Corporation, having been subsidized by WHDH television profits, would fold if we were allowed on the air, thus throwing 2,500 people out of work and leaving Boston with just one major newspaper.

In the middle of all of this, we were making final preparations to go on the air. Our days were filled with a flurry of interviewing, hiring conferences, and discussions on salaries and contracts. On March 14, we filed our response to Fortas's request for a stay: "Serious questions of public policy are presented by the assertion that an unprofitable newspaper must be kept alive from its profits from a television station in order to compete with other newspapers lacking such a financial crutch." The Court would have to act within three days, by Friday night, if their decision was to precede our going on air Sunday at 3 AM, for they did not convene on weekends. Meanwhile, I had already mailed out celebration invitations, which read, in part:

BBI has finally realized its "impossible dream." After two weeks of taping we are set to go on air Sunday, March 19th. It is time to celebrate . . . at the Marriott Hotel in Newton, Massachusetts, at midnight. Following refreshments and a buffet dinner we will leave for WCVB at TV Place about 2:15 AM. At 3:05 AM, WCVB-TV will make its inaugural broadcast and after that will leave its call letters and logo on air until the regular broadcasting day begins at 6:58 AM Sunday morning. . . . Long live WCVB-TV!

There were still technical glitches to fix. Some of our equipment wasn't working well and we were having trouble taping our first programs: sometimes the tape machines worked; sometimes they didn't. The door in the main studio wasn't wide enough for the big-camera dollies; the switching equipment was erratic, and the gates on the 16-millimeter film projectors kept popping open at inconvenient times. Because we were using off-duty engineers from WHDH to rehearse for a few hours at a time and they were unfamiliar with our new equipment the results were less than professional. As if that weren't bad enough, our taped 3 AM opening show wasn't all that good, but we had to go on the air with it—warts and all.

A Gut-Wrenching Day

On the morning of Friday, March 17, the senior staff at the station—all nine of us—assembled in my office to await the Supreme Court's decision. I sat behind my desk, while the others perched on the sofa or in chairs. The more nervous staffers paced the floor. A little before noon, I called Benny Gaguine. He promised to call the second he heard anything. But I couldn't wait—I called him again about an hour later. Then a little after that, Don Ward called to tell us that, according to the Clerk of the Supreme Court, nothing had happened yet; and no one knew when the justices would get to our case. In an effort to distract ourselves, we turned to guessing games, but when everyone was asked to write down the names of all 50 states, the best anyone could do was 48. At 2 PM, there was still no word. "My stomach is tied in knots," Bob Bennett announced and left the room to unwind. Off and on, everyone did the same.

I called Washington again an hour later. The clerk was still saying he had no idea what would happen; maybe they wouldn't even get to our case today. At this point I about lost it. "We plan on going on the air at 3 AM Sunday," I almost yelled, "and here it is Friday afternoon, it's starting to get dark, people are calling, the staff wants to know, we've arranged for the WHDH staff to come over." Don and Benny tried to calm me down. Nothing could be done to hurry the Court, so the best thing we could do was stay as cool as possible.

But we could hardly see or think straight. We gulped down more coffee. Night was falling; the lights went on. Holding our heads in our hands, we kept asking ourselves, will we go on the air or won't we?"

At 5:15, the phone rang again and everyone jumped. On the other end, a subdued Don Ward said we could give up for the night. The justices had gone home, and nothing was likely to happen on the weekend unless the clerk worked overtime to issue the Court's decision. "Are we going on the air Sunday morning or not?" I pressed him. The eternal optimist, Don said we should plan to and if we didn't hear anything to the contrary by then, simply go ahead.

Everyone got up and left the room, dejected. I stayed back alone.

Then the phone rang again. Don could hardly contain his excitement. "Leo," he shouted, "you're on the air! We just heard from the Clerk of the Court! The Court has denied WHDH's appeal for a stay, unanimously, except for the Chief Justice who recused himself." My heart was in my mouth, but I managed to thank him before hanging up and rushing out the door in search of someone—anyone—to tell. The people who'd been in my office all day had already gone, some of them probably to a bar to drown their sorrows. I strode toward the studios and the projection room and shouted at the top of my voice, "We're on the air!" General jubilation broke out. We called those who had left, and they hurried back to join in the celebration. At 5:39 PM, a UPI bulletin scrolled out of our Teletype: "The U.S. Supreme Court today refused to delay the transfer of a television station license which will require WHDH-TV in Boston to leave the air this Sunday. Channel 5 as of this Sunday will be turned over to Boston Broadcasters, Inc. The new station will use the call letters WCVB-TV."

We could now go ahead with the celebration dinner. On Saturday evening, with everyone seated in the dining room at the Marriott Hotel, and with smiles all around, I thanked all who had helped us: stockholders, lawyers, bankers, wives. I gave a special thank-you to Nathan David, whose hard work and selfless commitment to the larger good were key to getting us through the whole ordeal. I also singled out Bob Bennett, Dick Burdick, and Dan Prigmore, and ended by introducing our outside partners, First National Bank, our national sales representative, and our public affairs firm. Bill Poorvu and Matty Brown followed with remarks that could now—finally—draw on some humor to relieve years of strain and anxiety.

On the Air—Officially

About 2:30 AM, we left for 5 TV Place to watch WCVB go on the air. Seconds before 3 AM, our logo appeared and we all cheered and clapped. Then came a sort of New England scene, a lake with a canoe on it and the "5" in one corner. Ken Stall came on with a commentary about Channel 5. Matty Brown and I each spoke for a few minutes. Regular programming got under way in a few hours; the next day, newspaper critics wrote that there had been surprisingly few hiccups for an inaugural broadcast.

So we settled in for the long haul. Some months later, our treasurer, Bill Poorvu, reported that Boston Broadcasters had spent approximately $5.3 million on fixed assets (equipment, buildings, and improvements) and had incurred pre-opening expenses of about $2.5 million, not counting depreciation. We had raised $1.8 million from the stockholders, borrowed $5 million from the First National Bank of Boston and $300,000 from Harrington, Righter & Parsons, and expected to obtain $1.4 million through a mortgage on BBI's real estate holdings. This left less than $1 million for operating expenses, but we made do by stretching payments. Personally, I risked $222,552, plus all of my Bolt Beranek & Newman stock until the bank note was paid off. Others made similar commitments.

One of our stockholders, Stan Deland, a senior partner in the Boston law firm of Sherburne, Powers & Needham, sent me a note that I cherish, expressing his

admiration of leadership that you have devoted to what I hope will prove the most difficult period in the history of BBI. Your patience and tact set a shining example through times of stress and what might have otherwise been a storm, and your outward calmness lent an air of confidence during those very trying days. We are all looking forward to the development of what I am sure under your leadership will prove to be the outstanding television station in the country.

WCVB's news programs soon became the most watched of Boston's three TV stations—the other two were WBZ and WNAC (since 2000, named "WHDH"). Accordingly, our 30-second commercial rate during newscasts was higher than that of the other stations.

When it came time to apply for a three-year license, we documented our activities in great detail. By 1974, our news programming had increased to 15 hours per week and the locally produced, nonnews programs remained about constant, at 37 hours per week. We took special pride in our editorials, broadcast nightly five nights a week with trenchant commentary on pressing events and issues of the day. The FCC awarded us a normal, three-year license without comment.

Under the headline "Renaissance Man: Leo Beranek of BBI," in the September 29, 1975 issue of *Broadcasting,* a full-page "Profile" described me as "scientist, educator, businessman and now broadcaster—President of Boston Broadcasters, Inc., the latest in an array of careers he has pursued." The encomium to the 1976 Abe Lincoln Television Award conferred on me by the Southern Baptist Radio and Television Commission proclaimed that my "life as citizen and broadcaster exemplifies the ideals expressed by Lincoln, 'Firmness in the right as God gives us to see the right,'" and honored me "for superior community service programs and courageous editorials which have established WCVB-TV as a station which takes seriously its commitment to serve the public interest."

Under the headline "Some Say This Is America's Best TV Station," a full-page 1981 article in the *New York Times* focused on WCVB's innovative local programming. By this time, Bob Bennett had succeeded me as president and was becoming known in many circles as the best local broadcasting executive in the nation.

I served as president and chief executive officer of Boston Broadcasters from January 1963 until September 1979, my automatic retirement date at age 65. I became chairman of the board in January 1980, upon Matthew Brown's automatic retirement at age 75, and remained half-time in that post until 1983. WCVB-TV was sold to Metromedia on May 17, 1982, for $225 million, the highest price ever paid for a broadcast station until then. Our gamble had paid off.

9 Family, Nonprofits, and Variety

Quite apart from pursuing our normal professional or financial objectives, leading our everyday lives, and raising our children, most of us dip—and occasionally plunge—into ventures on the periphery of our primary sphere of experience, sometimes altogether foreign to it. In this chapter, I recount a number of such ventures in my own life, most of which belong to the period after World War II.

The birth of my sons brought new responsibilities and countless pleasures and rewards as I watched them grow up. I could now find time to take greater part in Boston's rich cultural life—the Boston Symphony Orchestra, in particular—as a devotee of music, administrator, volunteer, and patron. A deep personal loss—the death of my beloved wife, Phyllis—was followed by a second wonderful marriage, with a fresh new round of activities that included sailing, gardening, and traveling. I served as president of the American Academy of Arts and Sciences, and, following that, as a member of Harvard's Board of Overseers. A new business venture kept me energetic and on the ball, with ties—bridges, even—to a professional life that, although my age was advancing, I wasn't quite ready to leave behind. All this preceded my final professional commitment: helping to create a series of fine musical performing spaces in Japan—that story told in the next (and final) chapter.

On the Domestic Front

When Phyllis and I married in 1943, we moved into an apartment building on the edge of Cambridge Common, at 50 Follen Street. She worked as a dental hygienist until our first-born James (Jamie) arrived in July 1947, two months before I joined the MIT faculty. Our apartment was luxurious enough, we thought, for a pair of hardworking newlyweds; it boasted a large living room, large bedroom, kitchen, bath, and entry hall. Because it was wartime, we had no car and moved around on bicycles instead. Food was rationed, particularly meat and fish, so we ate at restaurants once or twice a week. We joined the Theater Guild and had a season subscription to the Boston Symphony Orchestra. I traveled on business several days each month, and when I was away, Phyllis often stayed with her parents in Jamaica Plain. About once a month, we had guests over for dinner, usually my Harvard colleagues or our neighbors.

Not long after our return from Buenos Aires in 1949, I was visiting my close friend, Smitty Stevens, a psychologist and professor at Harvard, when he took me outside to see his new Cadillac. He had purchased it a few days earlier at a large discount from a dealer near Central Square in Cambridge. Phyllis and I decided it was about time for us to have a car, too, so we got on the subway and headed to Central Square. When I told the dealer I knew Smitty Stevens and wanted a car like the one he'd just bought and at a similar bargain price, sure enough, I got one—a Cadillac for $2,400, decked out with luxury whitewall tires.

Suddenly, we were free to go out more. I recall an evening driving President Killian and one of the MIT Deans from a party back to their apartments on Memorial Drive, and hearing them in the back seat muttering that this was the kind of lifestyle a person gets to lead with an income from consulting. I can also remember lugging uncounted buckets of water down to Follen Street to spruce up this prized possession.

Life became busier after we moved to Winchester, the day before Christmas 1950. Our rented house was at 7 Ledgewood, at the end of a private road on which there were only four houses; a short section of dirt road separated us from another group of four homes. There was snow to shovel from our large circular driveway and from the sidewalks around the house. Furniture had to be bought—our choice was contemporary, we weren't much taken with antiques—and paintings framed and hung. Phyllis's biggest task was making curtains for the many windows, which she did in record time so that we could throw a house warming party. I had to tend to new needs— buying wood for the fireplace, taking out ashes every day, and seeing that missing slate shingles were replaced and pealing walls scraped, sanded, and repainted. Our seven neighbors were wonderful, all graduates of Harvard, and we visited each other's homes regularly. Commuting now took time—we were no longer a five-minute subway ride from MIT. I carpooled on alternate days with Richard Frazier, a professor who lived a few blocks from us, so that our wives could have the family cars on alternate days. Like me, Richard combined a career in education with consultancies for various labs and firms; we were members of the same MIT department—Electrical Engineering—and his expertise lay in electric circuitry and electromagnetic devices and systems.

Each spring, the two heavily wooded acres around our house, with their dozens of mountain laurels, azaleas, and rhododendrons, their scotch broom, crab apple trees, and blueberries, would burst into bloom. Phyllis and I loved to saunter along the rough-hewn paths. The place was just about perfect for us, and when the fall of 1952 came around, we decided to buy it.

The lawyer for Mrs. Parker, the owner, quoted us a price of $55,000, equaling, he said, what the Parkers had invested in the two acres and in building the house. But there was no way we could afford to pay that much: Massachusetts real estate rules at the time required a down payment of 50 percent and my savings amounted to less than $20,000. So I offered $35,000, which the lawyer dismissed as outrageously small. Unwilling to give up so fast, I went to his office in Boston to see if I could bargain with him. I said that was all I could afford, adding that when we first moved in, Mrs. Parker had promised to give us special consideration if she decided to sell. I also told him—somewhat presumptuously, I suppose—that the house was not worth more than what I had offered. The lawyer then said he would have the property appraised, with the final price placed either at $55,000 or the appraised value, whichever was lower. He gave me a choice of two appraisers: the Cambridge Trust Company or the Boston Five Cents Savings Bank. I chose Cambridge Trust because the president, Gardner Bradley, happened to be the brother of our next-door neighbor, Dudley Bradley.

One rainy Sunday morning, a dour appraiser from Cambridge Trust arrived at our door, opining as he entered: "Modern, I hate it." Naturally, I had no interest in extolling the house's many virtues. Instead, I pointed out all the negatives—no basement, wall-to-wall carpeting instead of hardwood floors, no lawn—and omitted mention of how much we loved it all the same. A few days later the appraisal came in at $37,000. I rushed back to the lawyer, almost beside myself with glee. With kitchen appliances, garden tools, and the like added in, the final price came to $38,500. We closed the deal in November 1952.

Ventures in the Arts and Nonprofit Management

Whatever financial worries we faced early on had needed as my income from Bolt Beranek & Newman grew. Soon Phyllis was able to have her own car—a handsome Buick—and she could take greater part in Boston affairs, particularly volunteer work at Trinity Church. My love for symphonic music went back to my role as timpanist in the Cornell College Symphony Orchestra in Iowa and, for a brief period, in the Harvard-Radcliffe Orchestra. Phyllis and I regularly attended concerts of the Boston Symphony, first in Sanders Theatre and later as season subscribers at Symphony Hall. We went out to Tanglewood often to enjoy concerts by the Boston Symphony Orchestra and by students at the Tanglewood Music Center. It became quite clear to me that one of the reasons Boston was able to attract and retain so many top-level professionals was the presence of a world-class symphonic orchestra and an absolute gem of a concert hall.

After BBN's success with the installation of acoustical panels in the Tanglewood Music Shed in 1959—for which I was mostly responsible—I made friends with many in the Boston Symphony Orchestra family, particularly within the management. In October 1968, the president of the board of trustees, Henry Cabot, invited me to become a charter member of a new board of overseers. Nineteen of us appeared at the inaugural meeting in Symphony Hall. We were told that the overseers were to focus on building a stronger financial base for the orchestra. In the prior fiscal year, we learned, the orchestra had run up an operating deficit of about $300,000—a year later the deficit nearly doubled, and the BSO endowment was only $8 million. The organization was struggling, in dire need of an infusion of capital and financial know-how.

I attended the twice-yearly overseer meetings, served on two committees, and made my voice heard on money matters over the course of the next decade. In March 1977, I got a letter from the chairman of the board of overseers, David Ives, which read in part: "There is unanimous approval—indeed acclaim—for the suggestion that the best candidate to succeed me would be Dr. Leo Beranek. . . . Your managerial ability is needed. . . . So you must accept." I did, and, as seems to be the case in just about every venture I have been involved with, I went right to work on bringing about change. Before my chairmanship, the trustees and overseers had had a number of separate committees, often with the same function but with efforts not well coordinated.

With the cooperation of the trustees, I managed to merge the various commit-tees so as to avoid duplication and to make our work more effective. In 1978, I originated the "Trustees-Overseers Handbook," which contained a short his-tory of the orchestra, the names, purposes, and membership of twenty-six trustee-overseer committees, and short biographies of each trustee and over-seer. Published annually, and one of my finest accomplishments in working with the orchestra, the handbook became very popular among members of the boards. The concept was soon adopted by other nonprofit organizations in Boston, and gradually embraced by similar organizations elsewhere.

In an attempt to tackle the orchestra's financial hole, a "Boston Symphony Orchestra Hundredth Anniversary Fund" was inaugurated in 1976, which gave us five years until the celebration of the centennial (the orchestra had been founded in 1881). By August 1979, two years into my chairmanship of the overseers, a total of $10.4 million had been raised, toward a goal of $15.7 million. The endowment-raising effort was moving well along when the chairman of the fund, Roderick MacDougall, asked to be relieved. This coincided with my reaching the automatic retirement age of 65 at WCVB-TV. My half-time work there as chairman of the board left me time to spare for other activities or for additional commitments to current ones, including the orchestra.

The BSO's monthly newsletter for December 1979 reported that the orga-nization's annual deficit, even taking into account the income from the $10 million endowment, was more than $1 million. This trend, we knew, could not be sustained and might ultimately exhaust the BSO's capital funds and spell the end of the orchestra. A desperate situation, to put it mildly. The newsletter announced:

President Nelson J. Darling Jr. has charged two trustees with the task of leading the BSO out of the dark in the crucial area of fundraising. . . . Beranek, as head of the newly formed Resources Committee, will oversee all fund raising activities. Mrs. [Jane] Bradley is chairman of the symphony's Centennial Fund. . . . The task is not an easy one. As a pair, Beranek's and Bradley's determination to get the job done seems evenly matched, although their styles are quite different: Beranek's is precise, thoughtful, and deliber-ate, carefully inscribing the side of a paper cup with contrasting figures to illustrate a point. With his methodical manner of speaking as he meticulously searches for just the right phrase, he is almost professorial. Mrs. Bradley moves quickly, almost impatiently. She is exuberant, even a touch theatrical as she pounds the table to make a point. . . . Both have come to believe in the necessity for a change in attitude on behalf of all those involved in fundraising activities, which means overcoming the traditional Bostonian reticence to discuss—perish the thought—m-o-n-e-y.

The "BSO-100 Fund" board had a half dozen wealthy members, a number of whom were officers in high-tech companies in the area. To make them aware (or more aware) of the importance of a healthy BSO to the vitality of Boston—and ultimately to their own commercial and cultural interests, I put together a set of flip charts displaying income and expenses for a presentation to the board of trustees. "The orchestra is in a financial crisis," I said, "and we must raise $1 million in spendable money this year. Four areas will have to contribute toward this: management must reduce its annual operating budget by $300,000; the fund drive must produce $6 million in new pledges in the coming year, which will increase our endowment income by $225,000; the trustees and overseers must increase their annual gifts from last year's $55,000 to $230,000, an increase of $175,000; and the Friends [of the Symphony] and Projects must raise $300,000 more."

This ambitious program was approved by the board of trustees. The trustees collectively pledged to increase their endowment gifts by $1.6 million. Donations began to come in from Jane Bradley's efforts as well, and annual gifts to the orchestra picked up overall. By the end of August 1980, just seven months after my appointment as coleader of fundraising, the BSO-100 Fund had brought in an additional $4 million dollars. Flush with success, we upped our goal from $15.7 million to $18.7 million, which we planned to achieve in time for the Centennial Celebration in October 1981.

I asked a number of corporate leaders to meet with me in Symphony Hall to discuss how they could help to further alleviate our crisis. Ray Stata and J. P. Barger, leading Boston high-tech executives, promised to help. A surprise offer came from someone completely unknown to Nelson Darling and me: Harvey (Chet) Krentzman, a Boston financier of small businesses.

By October 1981, the BSO-100 campaign had raised $20,150,000, far exceeding our original goals of $15.7 and $18.7 million. In addition, annual gifts in 1980–81 provided $1.7 million, net of all fundraising expenses. It had been a hectic, tumultuous effort. The most amazing result of this campaign was its impact on the annual deficits. From 1978 to 1981 the operating deficits and, thus, the annual withdrawals from the unrestricted endowment capital, had run between $1 million and $1.4 million. But in 1982, the operations made a net *contribution* of $150,000 to the endowment capital.

Chet Krentzman's proposal, a remarkably creative one, was that a special event—"Presidents at Pops"—be organized to raise a net of $500,000 from the business community. A combination of Pops concert and black-tie ball, the event would target top executives in Boston. The concert would feature one or more corporate personages, who would make cameo appearances on stage. Chet's plan called for selling 106 packages of 20 seats, each package priced at $4,000, to participating companies. I met with him on twenty-five planning occasions between the time he made his proposal, in November 1979, and the first Presidents at Pops concert in 1982. A big hit at the first concert was the appearance of the president of Bank of Boston, Richard Hill, as clarinet soloist, pumping out a single note at designated points in a Bernstein divertimento. For a quarter of a century since, Presidents at Pops has played to sellout audiences, and in its 25th year, 2006, it grossed $1.4 million. A second event, Christmas at Pops, was spun off to accommodate the many companies eager to get in on the excitement and publicity.

In 1983, I became chairman of the Boston Symphony Orchestra's board of trustees, a position I held for three years, after which I became honorary chairman. But this was far from a happy time.

On November 5, 1982, Phyllis died unexpectedly and I was left desolate. The memorial took place in Boston's Trinity Church, to which she had volunteered much time, and was attended by nearly 500 friends and community leaders. Executives from Bolt Beranek & Newman and WCVB-TV served as ushers. The rector of the church, Reverend Spencer Morgan Rice, led the service and delivered a beautiful encomium. Gifts in lieu of flowers enabled the church to buy a memorial silver chalice and a number of processional candlesticks. My sons stayed with me for a while afterward, and I was invited often to the homes of neighbors and several Boston families whom I had gotten to know over the years—all of them trying, not very successfully, to console me. I lost nearly 30 pounds, ending up at about the same weight I had been as a young graduate student.

But I pressed on regardless—partly to distract myself, partly as an outlet for restless energy—and kept finding new ways to occupy my time. Soon after my retirement as chairman of the BSO board of trustees in 1989, I was elected president of the American Academy of Arts and Sciences, an honorary society whose membership includes distinguished leaders from a broad range of fields and disciplines, such as mathematics, the sciences, humanities, medicine, economics, business, and the arts. I had been inducted into the academy with due pomp and ceremony many years earlier, in the fall of 1952, at the headquarters then located on Newbury Street in Boston. I recall Professor Harold Hazen of MIT, one of my sponsors, escorting me down a long, sloping aisle to the podium, where academy president Edwin Land was waiting to bestow membership on me. The academy had been founded in 1780 by John Adams, John Hancock, James Bowdoin, and other heroes of the American Revolution, to provide a place where scholars, educators, and political leaders could meet to hash out plans for the emerging nation. Over the decades, it evolved into an honorary society whose members, recognized for excellence or accomplishment, are elected annually by vote of existing members. I had attended meetings through the years. Serving in the 1980s on the Committee on Publications, the Nominating Committee, and the President's Advisory Committee, I was well known to members of the governing body, the council.

I got wind that I was being considered for the presidency sometime in 1988, when the Academy's executive director, Joel Orlen, said that the Nominating Committee wanted to talk with me. A few days later, he and Jerry Wiesner, past president of MIT and an active fellow of the academy since his own election in 1953, invited me for a chat in Wiesner's office. Joel, incidentally, had a strong MIT connection himself, having served for fifteen years as the provost's executive officer before joining the staff of the academy. Later, he told me that, having heard my answers to some of Wiesner's questions, he was certain I had Wiesner's support. And on December 28 the Nominating Committee asked me to serve as the next president, for a three-year term, and I accepted. By then, I had no other demanding activities either under way or in the pipeline, and I was confident I could commit plenty of time to the position.

The academy's headquarters are now in Cambridge on a hilltop over what is called "Norton's Woods." You enter the modern building, erected in 1981 as a gift from former president Edwin Land, founder of the Polaroid Corporation, through a series of imposing oak doors into a high-ceilinged atrium with a balcony on all four sides accessed by a stairway near the entrance. My office was just off the top of the stairway.

As the BSO's had been, the academy's finances were in precarious shape. The organization was losing nearly $300,000 a year with an endowment of $7 million. I worked with the editor of the academy's magazine, *Daedalus,* to cut production costs and to raise the subscription price. I made a modest cut in staff, and adopted an HMO-style health insurance plan, partially funded by staff contributions. Pleas for support from the membership brought an increase in annual contributions, and dues were increased. Finally, I launched an endowment drive that in four years netted $7 million. So pleased were the academy's fellows with these changes that they voted to extend my tenure two years.

Family, Old and New

On the personal side, I have had wonderful relations with my two sons, Jamie and Tom. When they were young, Phyllis and I would take them out of school to go skiing with us in Switzerland for a month each year. Making sure they brought along their school assignments, I would oversee their lessons each afternoon from 4 PM until dinner time, and would return the boys to the formal classroom after that month away in fine step with their classmates. Jamie earned his master's degree at the University of Iowa and has since worked part-time for the Iowa Historical Society. Tom graduated from Dartmouth College and went on to law school at the University of Chicago, where he finished as managing editor of the *Law Review.* I spend several days with them in Chicago each December, and they get to Boston once or twice a year. Outside of these family reunions, we keep in constant contact by telephone and e-mail.

My professional life—with all its ups, downs, pressures, and anxieties—has evolved side by side with a home life marked by extraordinary peace, stability, and joy. One of the happiest times in my life came with my marriage to Gabriella Sohn in 1985. In the fall of 1984, two years after Phyllis's death, I learned that the Sohns were about to divorce. When the divorce was granted, Gabriella and I began dating. We were discreet about our growing interest in each other because, under Massachusetts law at the time, a divorce was not final until a year after the decree was granted. In November, I attended a charitable auction at the American Repertory Theater in Cambridge, where I bought a trip for two to Rio de Janeiro. Gabriella and I arranged to fly there in January. Our reservations were for the Sheraton Hotel in Rio and we were to be picked up at the airport on our arrival. Gabriella left her car in my garage in Winchester, and I arranged for a young man—one of our neighbors—to house-sit while I was gone.

When we arrived at Logan Airport in Boston, we found that visas were required for entry to Brazil; because we didn't have them, we were not allowed to board. We decided to head for New York, get visas at the Brazilian Consulate the next day, and board a late afternoon flight for Rio. Unaware of our change in plans (I'd forgotten to contact them), the hotel sent a car to meet us. When we didn't show up, the hotel staff called the Sheraton headquarters in Boston, through whom I had made the reservation. The agent there, knowing that Gabriella was with me and that her ex-husband was president of Boston's largest travel agency, called him and asked if he knew why we had not arrived. Don started to piece things together. And then, as if that weren't bad enough, driving Gabriella's car to buy cigarettes late the following night, my young house sitter ran a red light and was stopped by a policeman. When the officer found that the car was registered to a travel agency, he suspected it was stolen and, to verify the young man's story, he called Don and asked him to come out to Winchester. By now, Don knew for sure that Gabriella and I had flown off together to Rio, although he handled the situation calmly, despite being yanked out of his bed after midnight. No sooner did Gabriella and I finally arrive in the lobby of the Rio Sheraton than we ran into the *Boston Globe*'s social reporter, who proceeded to announce our whereabouts in his *Globe* column the next day, ending wryly: "Now that's friendship, isn't it?"

We were married in August 1985, in the living room of the beautiful Blantyre House in Lenox, Massachusetts, with an Episcopal minister from Boston's Trinity Church officiating and the BSO's principal harpist, Ann Hobson Pilot, providing musical accompaniment.

Gabriella kept her house on Padanaram Harbor. When told I didn't know how to sail and that, being raised in the landlocked Midwest and finding life so busy after I moved east, I had never even considered the sport, she replied, "You are going to learn now. I am ordering a Beetlecat boat for you, and when it arrives, I'll get a sailing instructor to teach you how to handle it." This turned into a wonderful hobby: I sailed most summer weekends for the next eighteen years. Despite my friends' urging, however, I saw no reason to get a larger boat, sailing on my own as I did, and wanting only to make day trips. Absorbed in tending to her vast flower gardens, Gabriella had no interest in coming along.

Global Wanderings

Gabriella and I travel a good deal and especially enjoy our visits to New York City, with its unmatched energy and cultural vitality. Our trips to countries in both hemispheres have immeasurably deepened our appreciation of music, cultures, cuisines, lifestyles, and institutions around the world. One special trip, to Germany, comes immediately to mind.

It was May 1990. The Berlin Wall had fallen in November of the year before, and Germans found themselves swept up in a swirl of emotions ranging from excitement to anxiety. I was president of the American Academy of Arts and Sciences at the time. A letter arrived inviting me and Gabriella to come to Germany as guests of Chancellor Helmut Kohl, along with a dozen others, to take part in discussions on "Europe's Place in the World Economy: Prospects for German-American Relations and East-West German Integration." Also attending the colloquium were representatives from Coca-Cola, General Motors, Stanford University, University of California at Berkeley, Harvard, MIT, the *Wall Street Journal,* and the *Houston Chronicle,* as well as the American Academy of Arts and Sciences, California Academy of Sciences, the National Bureau of Economic Research, and the investment houses of Goldman Sachs and Kemper Financial.

Gabriella and I planned to arrive three days early, with the idea that we would poke around Cologne before going on to Bonn, capital city of the Federal Republic of Germany (soon to be supplanted by Berlin as the capital of a unified Germany). When we told the Chancellor's Office, they were delighted, offered to pay, besides our air transportation, for our stay in Cologne, and asked what we planned to do there. We flew first class on Lufthansa, arriving in Frankfurt the morning of Sunday, May 27, 1990. We were met by a graduate student from the University of Cologne who told us that she would be our guide throughout our stay, that a car would be at our disposal for all three days, and that she and the driver (both staying in our hotel) would tend to our every need, day or night. We boarded a comfortable train, also operated by Lufthansa, for the trip to Cologne. Through the picture windows of our private compartment, we marveled at the view as we raced north along the Rhine River, over steep hills, through lush valleys, and alongside well-tended farms.

We rested at our hotel, the Bristol, in the afternoon. That evening, our guide had tickets for us to a symphony concert in the old City Hall, converted to a concert hall. Our seats were good, but I asked if she could arrange with an usher to escort me after intermission to an unoccupied seat in another part of the hall, so that I could experience the acoustics from two vantage points. That evening we also learned, much to our delight, that tickets would be held for us at the Cologne Opera the following day.

I had made the chancellor's office aware that Gabriella was an expert on textiles and would like to know of any textile museums in or near Cologne. Efficient as always, our guide announced that we were to visit the famous textile museum in nearby Krefeld. The museum director met us at the door and gave us a private tour. He lunched with us and, when he learned we were at the Bristol Hotel and would be attending the opera that evening, he invited us for cocktails before the performance at his penthouse, a short distance away from the opera house, where we marveled at his large collection of rare prints and other art work. At the opera, we sat in the "president's box" along with six or eight others, apparently government officials.

The next day, we were given another private tour, this time of the outstanding art on display at the Roman-Germanic Museum. And, of course, we couldn't have gone to Cologne without strolling around the famous Cologne Cathedral, with its beautiful twin towers and stained glass windows. At that time of year, late spring, every restaurant featured "*Spargel*"—white asparagus—which we savored along with some excellent German wines. Our guide told us about student life in Germany and invited us to join a group of graduate students for a simple dinner that evening at one of their apartments. We had a grand time there talking about life in our two countries, about the politics and customs of Americans, and about our reaction to the fall of the Berlin Wall.

On the morning of Wednesday, May 30, we said good-bye to our guide and were driven by car to our hotel, on the Rhine just outside Bonn. Luxury of luxuries, each invitee had a car—thirteen chauffeured automobiles, in all, for our group. We were taken to the offices of the chancellor that very afternoon, while our wives went to see Beethoven's birthplace. We sat around a tremendous oval table. I was positioned directly opposite Chancellor Kohl. After introductions, he spoke for some 20 minutes and then asked each of us to give a 5-minute response. Germans, he said, had not forgotten the American contribution to their prosperity, in particular through the post–World War II Marshall Plan. He was impressed by how well President George H. W. Bush understood Europe. Germany was committed, he asserted with confidence, to keeping NATO intact so that the interests—indeed, the destiny—of the United States and Europe would remain intertwined. He looked forward to the formation of the European Community (EC) and anticipated that Germany would support it to the fullest. Great Britain would be reluctant to join, he feared, but at some point it would give in and seek shelter under "the big EC umbrella." He hoped for more give-and-take between our two countries—in educational, cultural, and other areas—and suggested the formation of a German-American institute to promote this.

When my turn came, I told of my own sense that the United States and Germany had been drifting apart for some years, and that one way to reverse this trend would be to increase the flow of exchange students across the Atlantic. I said that the American Academy of Arts and Sciences, and its journal *Daedalus,* stood ready to play a part in this; in fact, one entire issue of *Daedalus* the following winter was devoted to "Germany in Transition." In the evening, we dined at the home of Chancellor Kohl and his wife, Hannelore Kohl. We were served *Spargel,* and one guest mumbled that by this time he was *"ausgespargelt"* (asparagused out). As the oldest member of the delegation, I stood and gave the "thank-you" remarks.

Our focus the next morning was on Europe's place in the world economy. At noon, we joined up with Minister for Foreign Affairs Hans-Dietrich Genscher, who gave us his take on the fall of the Soviet Union and the Berlin Wall. In the afternoon, we met with the editors in chief of several German newspapers. That evening, we dined with the president of West Germany, Dr. Richard von Weizsäcker and his wife, Marianne Freifrau, at their residence on the banks of the Rhine. It was a beautiful late-spring evening—clear skies, with a cool breeze—and we had great fun chatting over cocktails, before dinner, on the open-air terrace. The next day, we flew to Berlin, where we met with representatives from East Germany for a discussion of what lay ahead for the combined Germanys. On the final day of our visit, we toured the Sanssouci Palace in Potsdam, former residence of Frederick the Great and the site where President Truman had met with European leaders to hash out the world's postwar future. Our day was capped off by an excellent luncheon with the mayor of Berlin, Walter Momper. In the evening, we attended a symphony concert in Philharmonie Hall and flew home the next day.

Harvard, Again

The first decade of my quasi-retirement was starting to look as busy as the peak decades of my professional career (1950–80)—if not busier. In the fall of 1983, I received a telephone call from the nominating committee for the board of overseers of Harvard University. Did I wish to have my name placed on the ballot to be sent to all Harvard alums? Five of the ten nominees would be elected, each for a six-year term. I thought for a moment, then said yes, I did, and wrote my formal letter of acceptance in December.

Later I wondered how on earth I could possibly be elected. Only two of the thirty overseers then in office were scientists or engineers, and most of the others held high positions in corporations, government, and universities. But considering my past activities, perhaps enough Harvard alumni would know me by name, if not by reputation. Qualified voters numbered 158,000, and 32,000 ballots were returned (20 percent, an even smaller turnout than in our national elections). As fourth-highest vote getter among the candidates, I was duly installed as an overseer. Among those on the board during my term were Peter Goldmark Jr., James Schlesinger, Landon Clay, Albert Gore Jr., Andrew Brimmer, Jerome Wiesner, John Whitehead, Saul Cohen, and Archbishop Desmond Tutu. I served on Harvard's visiting committees to the Physics Department, Biology Department, Loeb Drama Center, and Harvard Business School.

Vision, Innovation, and the Unpredictable Cycles of Entrepreneurship

In May 1980, Babson College invited me to a dinner held in connection with an awards ceremony honoring select entrepreneurs. I was early to the table and found An Wang (also early), whom I had met once before at a party in Winchester, seated next to my assigned place. I knew that An had invented a computer that engineers were finding very useful and that Wang Laboratories, which sold computers for businesses, was one of the most rapidly growing companies of New England. We struck up a conversation and somehow got onto the subject of the ARPANET, which Bolt Beranek & Newman had invented. I told him of our work in packet switching. Intrigued, he asked me to consider becoming a member of his board of directors.

On May 28, I was his guest for lunch in the executive dining room over at the Wang Laboratories. He introduced me to several other board members and we exchanged stories about our management and technical experiences. I said, now that I was working only half-time at Boston Broadcasters, I would have time to be on their board—which I joined on October 28, 1980. Faithfully attending monthly board meetings for the next six years, I brought in several new ideas on the management of research and on wide-area packet-switched networking, both carryovers from my BBN experiences.

At the time, Wang Laboratories aimed its computer sales almost entirely at secretaries and other office support staff. The principal product was a central computer, Model VS, connected by cables to terminals on desks. Secretaries could do data and word processing and spread sheets at these terminals with remarkable efficiency compared to a typewriter: text could easily be modified and blocks of information instantly interchanged within a single document or between two or more documents. The keyboard was standard; the operating system proprietary. Wang's business initially had almost no competition. In the company's 1979 and 1980 reports, for example, there was prominent mention of large companies that had adopted Wang word processing and other computing equipment in their secretarial departments—General Electric, Equitable Life, Pioneer, First National Bank of Chicago, Nestle, Finnair, Schlumberger, Rockwell International, American Express, Scotts, RCA, Goodrich, FMC, KLM, Jardines, and Willis, Faber & Dumas.

In August 1981, IBM introduced the "Personal Computer" (PC), for $1,995. It had a green monochrome display screen, standard keyboard, and could be used to do data and word processing and spread sheets, either as a stand-alone or networked computer; its operating system was MS DOS (Microsoft Digital Operating System). IBM's PC had several features absent from the Wang VS. For example, operators could make running checks for misspelled words on the PC; they could insert additional words without having to press the "insert" key, as they had to on the VS. Vital to its future marketability, IBM's PC could be connected to IBM mainframe computers using the so-called SNA network protocol.

In 1983, Wang Laboratories announced a personal computer with both MS DOS and CP/M operating systems. The June 30 company report disclosed that by the end of May in that fiscal year, 20,000 units were on order. Overall, annual sales rose 33 percent and profits were up 42 percent from the year before. A similar gain followed in 1984. Flush with optimism, An Wang got to work on yet another (his third) office building; investors snapped up the company's bond issues.

At this point, however, things began to go downhill. The computer experts around Boston had convinced An that IBM would be forced by international standards to adopt the X.25 protocol, which would make it possible for users of Wang's VS to connect their computers to IBM mainframes. But IBM bucked the international trend, and refused to allow connection to their mainframes via the X.25 protocol. Neither An nor anyone in his large research and development department nor I foresaw the impending disaster.

Our board meetings in the fall of 1983 were dominated by an offer from a company out west to buy Wang Laboratories at a high price. An was urged by half the members, including me, to accept the offer, but he envisaged only continued success for the company and wanted to leave his son in charge of a grand institution.

His enthusiasm grew unchecked the following year. "We have added to our leadership position in word processing," he wrote in his 1984 annual report, "and enhanced our data processing offerings to earn recognition as a major vendor. In the newer technologies of image, voice and networking, products like our Professional Image Computer (PIC), Digital Voice Exchange (DVX), WangNet and FastLAN give us technical superiority and the industry's most solid strategic development base from which to grow." Unfortunately, these products did not sell well.

By this time, most larger companies owned IBM mainframe computers for financial record keeping, personnel records, inventories, and sales information. Because this equipment was very expensive, complicated, and in need of constant attention, companies usually set up an information systems division, under the management of a high-level officer, to manage mainframe activities. As information systems managers increasingly sought to coordinate all computer activities within their companies, each found that the best (and simplest) way to do so was through a network of IBM personal computers connected to their central IBM mainframe. Because Wang's VS computers were not compatible with IBM mainframes, IBM began to corner the secretarial market as well. Wang also suffered because, in contrast with IBM, its sales pitch went primarily to heads of departments (who managed secretaries), rather than to the information systems executives. Frustrated salesmen kept reporting that information systems managers were refusing to purchase non-IBM products, even for dedicated secretarial use, and—worse—would routinely overrule any department head's request for a Wang system.

Without IBM mainframe connectibility, Wang computers became essentially unmarketable and sales to large corporations fell precipitously. Business spiraled downward early in 1985, with the company's June 30 report showing a free-fall drop in earnings from $210 million to $15 million. This was due to IBM's competition, in large part, but also to the growth of another competitor, Digital Equipment Corporation (DEC). An explained in the annual report that his company's reduced rate of growth—"tanking" would have been a more accurate descriptor—was the result of other companies cutting corners to maximize profits while compromising quality, and "a lack of adherence to industry communications standards." The research and development department undertook an intensive, last-ditch effort to render Wang products compatible with the IBM mainframe, which was successful—but too late.

An had already announced his intention to have his son, Fred, succeed him as company president. This was a frightening prospect: Fred had no executive experience to speak of—nor was he a strong or decisive character. The company desperately needed the services of a turnaround expert; a routine passing of the reins from father to son was a terrible idea. Four board members—Richard Smith, Peter Brooke, Louis Cabot, and I—met privately with An in October 1985, over lunch at the Union Club in Boston. We tried to talk him out of it, but he was adamant. We met with him twice more in the next few months, with the same result. In October 1986, Fred became president. By this time, in the absence of a shakeup in management and direction, the fate of Wang Laboratories was a foregone conclusion. IBM, Digital Equipment Corporation, and Hewlett-Packard took over the market. I stopped being a consultant in 1986 and resigned from the board that fall. Fred resigned as president in August 1989, the year Wang's annual report showed "earnings" at negative $424 million. Once widely heralded for its promise and achievement, the company was soon forced into bankruptcy.

A Healthy Life

It feels good to be alive and kicking at 93. My doctor says I have the body of a 75-year-old; my dentist tells me I have the mouth of someone half my age. What's my secret? People often ask. Genes have a lot to do with it, I'm convinced. Grandmother Beranek lived to 108. When she reached 105, a television reporter asked for *her* secret. "I have never been sick," she replied—a simple, direct response, though certainly not the litany of "magic bullets" the reporter and his audience were hoping to hear. She delivered food to folks stricken during the 1917–18 flu epidemic without becoming ill. Grandmother's own parents lived to well past 90.

The whys and wherefores of living to a grand old age, with body and mind reasonably intact, remain a mystery. I don't expect to match—much less exceed—grandmother's record, but I have tried to smooth the path to whatever final age is in store for me.

My efforts at healthy living started early on. I worked as a hired hand on an Iowa farm for two summers while I was in college, feeling rejuvenated each time and raring to take on the stresses of the approaching school year, which included fixing radios and playing in a dance band. I was never tempted to smoke, even though my parents had no objection, and I have never drunk much in the way of alcohol. I had my tonsils removed at age 33 following a period of chronic sore throats, compounded by a severe case of strep throat. At 46, I had my appendix removed and stayed away from work just three days. Until I turned 88, I skied on the most advanced downhill black trails at Alta, Utah. On my last trip there, every day-lift pass announced in bold letters: "Leo is 88 and still skiing. Great!" But, at 90, I suddenly found myself unable to walk normally. MRI and CAT scans and a myelogram revealed several pinched nerves in my lumbar vertebrae. At 4 PM one day, neurosurgeon Gerwin Neumann performed corrective surgery at the New England Baptist Hospital. At 1 PM the next day, I walked out of the hospital to our car, where Gabriella was waiting to take me home.

In my early thirties, I started taking a multivitamin tablet daily in the hope that this would delay the graying process—and, perhaps as a result, my hair didn't turn gray until I was almost 50. I continued to take multivitamins because I felt they contributed to my good health in a number of ways. I haven't had so much as a common cold, for example, in thirty years. When I underwent knee surgery in 1988, the doctor warned that arthritis would soon follow. I went to the library to read up on it. The medications discussed were only for relief on pain, I found, although all books urged exercise as a useful preventive. One book appealed to me because its outlook was relatively upbeat. The author urged the use of milk with two rules: Drink a glass of whole milk at each meal; do not drink any other liquid in the periods 15 minutes before or 2 hours after taking the milk. If not diluted, the argument went, the milk fat would head straight for the small intestine, avoiding the liver and instead ending up at the joints where it would do the most good. I have religiously followed this regimen for 28 years, and still have yet to feel the pangs of arthritis pain.

My daily food intake is highly disciplined. Each morning I prepare a bowl of oatmeal—the old-fashioned kind, not the instant variety—and top it with a sliced banana. Oats are a fine source of both fiber and protein. I eat an apple, pear, peach, or orange, usually twice a day. When I am at home for lunch, I whip together a salad with lettuce, tomato, cauliflower, broccoli, mushrooms, and carrots—and a sprinkling of feta cheese to add zest. For my evening meal,

I prefer chicken, low-fat pork, fish, or eggs, usually accompanied by a cooked vegetable and a baked potato or rice. I seldom eat bread.

Every morning before breakfast, I exercise, first at home and then at the nearby Wellbridge Health Center. At home, I spend 20 minutes doing stretching exercises for the arms, legs, and abdomen. At Wellbridge, I ride a stationary bicycle for 10 minutes, bumping my heartbeat up to 118. After that, I stride the treadmill for 20 minutes at over 3 miles an hour. Then I do a series of exercises, some with weights, designed to "lubricate" the joints most likely to be affected by arthritis—shoulders, hips, knees, and elbows. My fingers get enough of a workout on the PC keyboard. I take no medicines of any kind, not even pain killers.

I don't remember names as well as I used to, but I'm still pretty good at crossword puzzles and I read lots of books. My hearing has deteriorated to what's probably about average for my age, although my digital hearing aids are a tremendous help. I had been severely nearsighted for most of my life, but at age 77, with the onset of cataracts, lens implants gave me better vision than I ever had even in my best younger years. I'm still socially active and enjoy going to meetings, concerts, and parties—nearly always with my wife, Gabriella. In the fall of 2007, I have been invited and shall give major technical papers at the International Congress on Acoustics in Madrid and at the Audio Engineering Society Convention in New York.

The Acoustical Society of America met in Seattle in May 1988. I happened to be on a day cruise—one of the society's planned recreational events—when a Japanese acoustical engineer, Takayuki Hidaka, approached me. My book *Music, Acoustics, and Architecture* (Wiley, 1962) had been translated into Japanese in 1972, and Hidaka had studied it with great interest. He asked some questions about a paper I had presented that day, and then about my current consulting activities. Although I thought nothing of it at the time, a month later, he wrote to say that his boss was planning to invite me to consult on a new opera house project in Tokyo. Thus began twelve years of collaboration with the Takenaka Corporation and the project architects, TAK Associated Architects of Tokyo.

My first trip to Tokyo came in early March 1989. I arrived in the midst of a typhoon and, dampened, proceeded by bus to the Palace Hotel, a grand establishment across the street from the walled-in Imperial Palace grounds. My room was excellent and I had two nights to recover from jet lag before my first encounter with the acoustics group in Takenaka's Research and Development division.

On Monday morning, Hidaka met me in the lobby of the hotel. He was then about 35 years old, engaging and attentive to every detail. His style of dress tended toward the formal, a dark suit, white shirt, and conservative tie. Married with three children, he was a natural leader skilled at coaxing others into action. He spoke English reasonably well and, judging by the way he grasped written material, read it flawlessly.

The project for which I was brought on as "Acoustical Design Consultant" was the New National Theater's Opera House in Tokyo. It was to be built on a large plot, the site of an abandoned city water purifying plant, a half mile west of Tokyo's busy Shinjuku Station. Now I had a chance to apply in real life some of the knowledge gained from my long-term acoustical studies of concert halls. The Japanese government had held an international architectural competition for the project, which was won by Takahiko Yanagisawa, president of TAK Associated Architects. The New National Theater project had three components: the Opera House, the Drama Theater, and the Experimental Theater.

Opera House

After looking over the in-progress drawings and acoustical studies for the Opera House in Hidaka's laboratory, I told him I would propose a new concept to Yanagisawa—that the proscenium be surrounded with reflecting panels on the audience side. After lunch, we proceeded to the offices of TAK, where Yanagisawa and his team of four architects sat on one side of a large rectangular table, and the Takenaka acoustics group and I sat on the other. A cardboard model of the hall stood at one end; large drawings were spread out in the middle. An interpreter was there to provide simultaneous translation of our discussions.

After introductory remarks on the principal acoustical parameters for opera houses and their conventional layouts, I explained the most serious problem in any opera house—that the orchestra has greater strength, is basically louder, than the voices of individual singers. Singers often complain about being "drowned out," while instrumentalists in the pit grumble that they are not allowed to play in forte passages as loudly as they should. To solve this problem I proposed that the opera house embody what amounted to an acoustical horn, which I sketched out on a piece of paper, with the bell of the horn on the audience side of the proscenium. As Yanagisawa followed my presentation with keen attention, I went on. The hall must have a reverberation time, when fully occupied, of between 1.4 and 1.6 seconds, and the sound must be evenly distributed.

After flying home, I got separate letters from Hidaka and a member of his team, who wanted me to return in August. At that next meeting, I was given a report a half inch thick, prepared by Hidaka's research and development group, containing acoustical comparisons between typical opera house designs and my proposed "horn." The comparisons were made using computer simulations and actual small-scale models. The data showed that the "horn" design was better acoustically than the conventional design of an opera house, as I had predicted.

The meeting seemed to go well. Yanagisawa sent me the minutes, along with some queries. Some of his technicians were of the opinion, he wrote, that 2.0 seconds was the optimum reverberation time. He also reported how upset the lighting people were about my recommendation to remove one of the three lighting bridges in the ceiling—the one nearest the proscenium— shown on the original drawings. I stood my ground on both counts, noting that no acoustician I knew would disagree with my recommended reverbera- tion time of 1.4 to 1.6 seconds for an opera house, and that, although some opera houses such as La Scala in Milan have reverberation times as low as 1.2 seconds, in no important house did the time rise to as high as 2.0 seconds. Indeed, I went on to explain, sung words are virtually unintelligible if the reverberation time exceeds 1.6 seconds. I addressed the lighting issue as a matter of priorities, relative levels of give-and-take. Acoustics must take pre- cedence over lighting—visuals being more flexible than the delicate balances and nuances of sound—and because this was the first Western-style opera house in Japan, hence a model for future opera houses, I urged no watered- down compromises in design and construction.

I went back to Tokyo in November, by which time my "horn" design had been incorporated into the architectural drawings and model. The "horn" consists of a large reflecting canopy, which curves upward above the audience side of the proscenium, a second curved surface, which emerges just above the pit, extending out to and including the front edge of the first balcony, and the front ends of the upper two balconies, which are shaped to complete the "horn." A 1:10 scale model permitted Takenaka's acoustics group to work up a large amount of acoustical data. The data were positive, and the design looked eminently workable.

But after this third trip, to my great surprise, I heard nothing more from Tokyo for almost a year. Finally, in September 1990, I wrote to Yanagisawa ask- ing what had happened. Apologizing for not getting in touch, he announced that construction would be delayed for two years or so, to allow the govern- ment time to coordinate construction of the opera house with that of a con- cert hall to be built on the adjacent lot.

When the design meetings resumed, I was told of actions the Japanese gov- ernment had taken before announcing details of the original architectural competition. Among the several advisory committees it had created was one

on acoustics, which consisted of Japan's leading acousticians. These had prepared acoustical specifications for the official competition bulletin; Professor
Hideki Tachibana of the University of Tokyo had been granted funds to construct a model, which he had completed. Although the committee members
had assumed that either their committee as a whole or one of its members
would be appointed official project consultant for acoustics, after winning
the competition, Yanagisawa suggested to government officials that neither
Tachibana nor any member of the committee should be retained.

So there I was, an interloper perched next to a potential hornet's nest. Over
a year later, in January 1992, an official-looking fax arrived:

As the Construction Phase of the project approaches, we have selected you from a list
of international candidates, and request your advice and instructions concerning the
acoustic system of the theater. We invite you to come to Japan subject to the following
conditions: (1) Come to Japan 21 to 28 February, (2) Express your views concerning the
overall technical aspects and acoustical systems, (3) Exchange views with members of
the staff of the Agency for Cultural Affairs and the Ministry of Construction who are
involved in the project, (4) Fee of $12,000 including travel and expenses, (4) No deputy
will be accepted in your place.

And the fax bore the imprimatur of the Kanto Regional Construction Bureau,
Japan Ministry of Construction.

As soon as I accepted the bureau's invitation, the pace picked up. I flew
to Tokyo to make a presentation on Monday, February 24, to the Construction Preparatory Committee, a revamped committee for the opera house that
included one Japanese acoustician from the original committee, several specialists on theater equipment, and representatives of the relevant government
agencies. I went into considerable detail, including historical information on
opera house acoustics, attributes critical to acoustical success, and general
specifications. After I discussed the architectural plans presented to me, the
acoustician raised some polite questions and others asked me to expand on
certain points. The meeting ended on an upbeat note—we all seemed to be
in general agreement.

Meetings with government officials were scheduled for the next morning.
Yanagisawa and I went to the national government's complex of buildings,
not far from the Imperial Palace Grounds, visiting first the Ministry of Construction and then the Japan Art and Cultural Promotion Society. At each
meeting, we gave a shortened version of the preceding day's presentation.
An interpreter was on hand once again, and Yanagisawa and I answered

questions, which were rather perfunctory. The officials at both the ministry and the society seemed disengaged, so much so that one official actually fell asleep during the meeting.

At noon, the Ministry of Construction hosted a luncheon in its impressive dining room, with some twenty guests in attendance. Government leaders, Yanagisawa, and I sat at the head of a U-shaped table; the others surrounded us along the table's outer edges. I remember that the soup had flecks of gold in it—a custom, I was told, reserved for the most special occasions. After a delicious meal, various officials gave welcoming speeches in Japanese, interpreted for my benefit, and I responded with a speech of my own. I had a sense that the government fully supported my role as consultant and, when I returned home, Yanagisawa confirmed this by letter.

Half a year went by before I again received word. "The Second National Theater construction will start October 2, 1992," another fax informed me, "and the grand opening will occur in September 1997. The construction will involve a joint venture of five companies, of which the Takenaka Corporation will be the lead construction company." I went to Tokyo three times in the next two years to work with Hidaka's group, to meet with and answer the architect's questions, and to make sure that every acoustical detail was handled with care. I can truthfully say that the design embodied everything I wanted.

We ran our first acoustical tests in the Opera House as soon as it was completed in February 1997. With no audience, we had a tenor and a soprano sing, first to test for echoes—there were none—and for uniformity of sound from different parts of the stage. The results were entirely satisfactory. The Japan Shinsei Symphony Orchestra and singers under the baton of Kazunori Akiyama gave a special tuning performance with full audience on February 15. Nine excerpts were performed from operas by Beethoven, Wagner, Weber, Mascagni, Puccini, and Verdi.

Selected members of the audience—recording engineers and trusted longtime concertgoers chosen by Hidaka—were asked to complete a questionnaire on the hall's acoustical qualities. The result reflected a consensus that the reverberation was just right, the balance between the orchestra and singers was satisfactory, and the quality and volume of sound were preserved as singers moved about the stage. The singers reported that reflections from the walls of the seating areas were perfect, giving a sense of the hall as alive and responsive. The measured reverberation time with full occupancy was right in the center of the desired range—1.5 seconds at middle frequencies.

The inaugural production, Ikuma Dan's *Takeru*, was performed October 10 and 11. The Emperor and Empress arrived with much fanfare and sat in the front row of the first balcony. Prime Minister Hashimoto declared from the stage: "This opening of our new opera house marks Japan as a leading cultural center of the world." At the two performances, I sat in four different seats and found the sound to be satisfactory from each location. After the second night's performance, Hidaka and I interviewed the principal people involved for their reactions.

Toshiya Inagaki (sang the part of Takeru): It is very easy to sing in this opera house. It is just right. It is a great pleasure for me to sing here.

Ikuma Dan, librettist and composer: The sound was perfect—wonderful! The balance between singers and orchestra was just right. Every voice and every instrument were clear; the definition was just right, and everything went together as if one. Wonderful, I have heard operas in many theaters in the world, but this house is my most favorite one.

Yutaka Hoshide, conductor: The house is very wonderful. The singing was clearly heard in the pit, even when the orchestra played full strength.

Tsuneyoshi Kobayashi, chief producer: The acoustics are wonderful. The sounds of the orchestra are very much blended as heard in the audience. The conductor let the orchestra go full force and the orchestra blended with the singers, not covering their voices.

The volume levels, bass warmth, spaciousness, and clarity of sound, as measured in the audience areas using various electroacoustic instruments, matched the values found in the world's best opera houses.

Thus ended my most important commission, which has given Japan a performance space with optimal acoustical values. I have not heard a negative word about the acoustics since, and only one complaint in all—from the lighting designer about the elimination of one of his lighting bridges.

Concert Hall

The New National Theater was not my only project in Japan. There were six, of which the second was equally important: the Tokyo Opera City (TOC) Concert Hall, adjacent to the New National Theater complex, on a site that includes the TOC, a skyscraper, museum, galleria, shops, and restaurants—a mini cultural mecca of sorts. The two complexes cover eleven contiguous acres, a huge commitment of space in one of the world's most congested metropolitan centers. Yanagisawa was lead architect once again, although only for the Concert Hall part of the TOC complex. The design group included Takahiko Yanagisawa, Hiroshi Wada, Paul Baxter, Takashi Ninomiya, and Hiroshi Takemasa from TAK, and Takayuki Hidaka, Toshiyuki Okano, and Noriko Nishihara from Takenaka. TAC began designing the Tokyo Opera City Concert Hall in 1990 and retained me as an acoustical consultant in March 1991.

Yanagisawa had already decided on a number of features and had the technical specifications ready before I was invited in. The drawings showed the hall as a rectangular box 168 feet long, 66 feet wide, and 40 feet high. At one end of the floor was the stage, and the auditorium—with 1,630 seats in all—was positioned on the remainder of the floor beneath wraparound balconies at two levels. The ceiling was most unusual, a pyramid thrusting 51 feet upward from the rectangular box, its peak 91 feet above the auditorium floor. The interior surfaces were shown on the plans as heavy wood, and a spherical reflector hung above the stage.

The ceiling is difficult to visualize without actually seeing it. The peak of the pyramid is not directly above the center of the rectangular base, as one would normally expect. Imagine that someone has taken hold of the initially centered peak and pulled it, at the same height, to a position above the front of the performance stage. Of the pyramid's four triangle-shaped walls, the two side walls are identical, but the two end walls are quite different. One end rises steeply from the wall at rear of the stage to the peak; the other rises gently to the peak from the rear of the auditorium.

This novel feature was not to be found, that I knew of, in any other concert hall. I felt confident, however, that if sufficient volume were included, we could achieve the desired "singing" tone, and if the walls of the pyramid were made sufficiently diffusing and the balcony fronts below properly shaped, a comfortable feeling of acoustical spaciousness could be achieved. One problem here was the risk of a significant echo. But I hardly expected, both in length of time and effort, what the ensuing negotiations would require to make the details of the pyramidal ceiling both architecturally pleasing and acoustically effective. Visions of the Lincoln Center nightmare popped into mind.

I recommended a reverberation time of 1.9 seconds at full occupancy, the standard optimum measurement for musical performance. I specified a different design for the canopy above the stage and went over details of the stage structure with a fine-tooth comb. Because a relative small enclosure makes for fine ensemble sound and project, I recommended a stage area only slightly larger than that in Boston Symphony Hall.

By February 1992, the Takenaka acoustics group had constructed an accurate wooden model of the hall to scale (1:10). In it, we could place tiny listening heads (one-inch in diameter with two tiny microphones for ears) anywhere in the simulated audience seating areas and, on stage, position tiny loudspeakers to replicate the sound of musical instruments. Sophisticated computers were used to analyze the data. After studying a lengthy report prepared by the Takenaka group, I wrote:

> The mid-frequency reverberation time with full occupancy measured in the model is 1.9 seconds; the planned width and audience size yields satisfactory measured data on all other acoustical attributes. Our principal job now is to plan for adequate irregularities on the many surfaces in the hall in order to obtain a good patina and freedom from echoes. I recommend reducing the [hall's] volume slightly in order to achieve the desired reverberation time of 1.9 seconds.

At a second meeting in September 1992, I raised the problem of ceiling design. The largest surface of the pyramidal ceiling, I stated, was so sloped as to send a treacherous echo back to the orchestra. I warned that the echo must be reduced either by changing the shape of the surface or by adding substantial irregularities; otherwise, the risk of an acoustical calamity was quite high. I added my reservations about details in the design of the reflector above the stage. When the architect asked me whether the reverberation time shouldn't be set at 2.0 seconds, I said again—as I had repeatedly—that the sound quality would be better if it were 1.9 seconds. I urged, too, that the acoustical reflector above the stage be curved to resemble the horn-shaped design over the proscenium in the Opera House. Yanagisawa protested that his design called for straight lines, and that curves would be impossible without scrapping the design and starting all over again. If there had to be straight lines, I said, then the canopy should be a large solid body with a flat, irregular surface underneath, which would mean abandoning the notion of a sphere over the stage.

As time went on, it became more and more clear that the long pyramidal ceiling would remain the most difficult, contentious design feature. Yanagisawa wanted a simple, smooth surface for all four ceiling surfaces; I countered that such would be an acoustical disaster, producing a major echo from the largest surface. We continued to debate the design of the stage canopy. And when we met again two months later, I kept insisting that the stage canopy needed to be large and its undersurface irregular. What was most critical, I said, was that we needed to add irregularities to all the surfaces of the pyramidal ceiling to solve the echo problem and to give the sound a better patina. Yanagisawa, who had been calm if firm to that point, finally lost his temper. "We must not have a hall that looks acoustic," he exploded, "it must have a good architectural appearance—your recommendations tend toward a contrived interior and I cannot go along with them!" Was failure in sight?

In mid-November, I was asked to go to Tokyo to report, individually, to the Urban Planning & Design Institute, the Joint Tokyo Opera City Design Team (this group was responsible for all the buildings and construction at TOC, not just the concert hall), and TOC's nine sponsor companies. About twenty people attended each of these meetings. I did not mention the struggles between the architect and me, focusing instead on the characteristics that I was confident a concert hall must possess to meet expectations. I concluded, in fine diplomatic fashion, that our team had worked together smoothly and achieved an architectural and acoustical blend, although some details still remained to be ironed out.

When I met with the architects after my presentation, however, I threw diplomacy to the winds—desperate as I was to save the project from almost certain acoustical catastrophe. We must first solve the shape of the undersurface of the stage canopy, I kept pleading, and in the pressure of the moment I conjured up a neat illustration of the problem. I took out a large handkerchief and held two corners, while I asked Yanagisawa to hold the other two. "There," I exclaimed, "see how the center sags down. Divide the under surface of the canopy into squares, say, 2 to 3 feet on a side; in half the squares make the bulge push downward, and in the other half make them push upward." The stratagem worked, showing how effective a simple graphic can work—more effectively, to be sure, than any amount of droning on in technical jargon, especially where language is a barrier. My recommendation was accepted.

In June 1994, the Takenaka acoustics staff presented a thick report of measurements made in the 1:10 scale model. Several stage canopy designs and slopes had been measured for acoustical effectiveness, with excellent results. The best was a square canopy, 24 feet on a side, tilted 8 degrees from horizontal. The uniformity of sound reflections throughout the hall equaled the best I had seen for a hall anywhere in the world—at every seat, the first reflections arrived within 20 milliseconds after the sound that arrived from the stage. Further, measurements showed over 10 early reflections from balcony fronts and the side walls in the first 80 milliseconds after the direct sound as measured at all seats. This number equaled that measured in the Vienna Musikvereinssaal, an acoustical gem.

The problem of echo from the pyramidal ceiling, however, remained unsolved into 1995, even though we kept exchanging dozens of faxes. Meanwhile, at the request of the owners of the TOC hall, I met in Boston with the Boston Symphony Orchestra's music director, Serji Ozawa, to ask if he would conduct the opening concert on September 10, 1997. He agreed. Sometime later, I asked if he would conduct the tuning concert scheduled for June 11, 1997, to which he also agreed. We were very lucky to have such an internationally famous conductor—himself a Japanese icon—take charge of these two critical concerts.

The controversy over the echo problem climaxed on August 8, 1995, at a special meeting in the architect's office. In attendance were Yanagisawa, Wada, Baxter, Ninomiya, Matsuzawa, Hidaka, Masuda, Nishihara, and myself. I had already specified a new kind of sound-diffusing surface on the rear wall of the drama auditorium for the New National Theater building across the way (about which more later). In that hall, the rear wall is circular with its origin at the front of the stage. Untreated, this configuration would produce an intense echo in the front of the theater. The treatment I had proposed was a structure known technically as "quadratic residue diffuser," QRD for short, that covered the entire circular wall, about a foot deep. To make the unattractive QRD acceptable acoustically and visually, it was covered with narrow slats spaced so that the open area exceeded 50 percent. Hidaka, of course, had worked with me on that Drama Theater design and he now wondered whether we should also use QRDs in the Concert Hall. I had not seriously considered them because of their awkward appearance, and I knew that a ceiling surface entirely covered with slats would be visually unacceptable. Yanagisawa then offered an ingenious solution to the appearance problem: first, break up all four surfaces of the ceiling into a series of steps, each about 20 inches high; then, on the long controversial rear sloping portion of the ceiling, provide each step with a QRD. Instead of covering the QRDs with slats like those used in the drama theater, he said, each could be covered with a mesh screen which would be transparent to sound, but not offensive to the eye. The next day Hidaka and colleagues began testing the proposed QRD for each step. They reported that the design was satisfactory, and it was immediately incorporated into the plans. What a relief!

The hall was completed in June 1996, with all finishing touches in place. The first acoustical measurements, made in July, gave us confidence that it would be a resounding success.

Maestro Ozawa, on a routine visit to his home country, came to the hall unannounced on October 10. He heard no music, but made negative remarks that upset the hall manager, expressing his amazement that I had agreed to this risky pyramidal ceiling. When I met with Ozawa in Boston, I showed him the impressive acoustical data, which placed the TOC Concert Hall among the very best. But, with the disastrous opening concert in Lincoln Center in 1962 in the back of my mind, I urged him—a bit forwardly, he may have thought—to use a normal-sized orchestra and not to attempt maximum sound in the relatively small auditorium. I also asked him not to extend the stage, as optimal sound would emerge using its normal dimensions. (Once again—shades of Lincoln Center, 1962.)

As we continued with our acoustical measurements in the hall, the Takenaka group and I felt that, even with orchestra, organ, and trial audience present, some additional acoustical material would be needed to achieve the desired 1.9 seconds. We determined that a suitable place would be on top of the light-reflecting tray just below the pyramidal ceiling, with glass wool about 3/4 inch thick, to be installed before the Ozawa tuning concert. Concerned about cost, Yanagisawa asked if it would be possible easily to remove the material if it proved unnecessary. Assuming that the glass wool would simply be laid in place, I assured him it would. Orders went out for installation.

On Wednesday, June 11, 1997, Ozawa led the New Japan Philharmonic Orchestra (Shin-Nihon Philharmonic) in the tuning concert with full audience. The program consisted of three certain crowd-pleasers: Brahms's Symphony no. 3 in F major, Tchaikovsky's Piano Concerto no. 1 in B-flat minor, with Momo Kodama as soloist, and Brahms's Alto Rhapsody, with Hanna Schwarz as soloist. The audience raved. I listened from two different vantage points, however, and was not happy with some of what I heard: in particular, a foggy acoustical sheen clouding the middle frequencies. Nobody else observed this, but my critique was taken seriously and Hidaka and crew performed a new battery of acoustical tests. When these proved inconclusive, we found ourselves at an impasse on what to do.

A short time later, after deep thought, I concluded that the audience was unexpectedly absorbing more of the mid-frequency sound with the pyramidal ceiling than would have been the case with a flat ceiling. Thus the glass wool that we had added was a mistake; because it also absorbed more sound at middle than at low frequencies. I recommended its removal. This request was not at all well received because the contractor, instead of laying the glass wool in place, had screwed a perforated (sound-transparent) covering over it. Now, instead of removing this covering and the glass wool, an even more expensive alternative was taken—a solid covering was screwed over the perforated one to nullify the sound absorption.

The opening concert took place on Wednesday, September 10, 1997, with Ozawa once again on the podium, conducting this time the Saito Kinen Orchestra. The lone piece performed was Bach's massive *Saint Matthew Passion,* with John Mark Ainsley as the Evangelist, vocal soloists Christiane Oelze (soprano), Nathalie Stutzmann (contralto), Stanford Olsen (tenor), Michael Volle (baritone), and Thomas Quasthoff (bass-baritone), and with the Tokyo Opera Singers and SKF Matsumoto Children's Chorus. The most important event in the history of any concert hall, arguably, is the opening night's performance. We were most fortunate, too, that Maestro Ozawa had selected a composition that would test just about every acoustical aspect of the hall. The orchestra was of average size, avoiding any need for extension of the stage. The vocal soloists, all European specialists in Bach's choral music, were among the best in the world, and they were not pushed to their vocal limits.

The sound, overall, was glorious. I could detect no acoustical faults, and neither could any of the orchestra members, architectural and acoustical staff, or anyone else who offered an opinion. Ozawa said the hall was very good. The concertmaster (an American) praised the sound and observed that the instrumentalists had heard each other well.

The next afternoon, the Tokyo Opera City project personnel met in the architect's office. In attendance, besides me, were Yanagisawa, Wada, and Baxter of TAK and Masuda, Hidaka, and Nishihara of Takenaka. Yanagisawa positively glowed about the wonderful acoustics and noted the enthusiasm of performers and listeners alike. I agreed, observing that, in my view, the *Saint Matthew Passion* fitted the hall ideally for balance, projection, and warmth, and that the concert would go down in history as one of the best of Ozawa's career. The reverberation time, I added, measured 1.95 seconds with the sound-absorbing material covered over, and the result was spectacular. I had attended concerts in other Japanese halls—Suntory, Orchard, Tokyo Metropolitan Arts Center, Bunka Kaikan, and Kioi Halls—and the sound in them paled by comparison. I finished by praising the TOC Concert Hall as "architecturally splendid," "a thing of remarkable beauty," which pleased Yanagisawa.

And the hall has since that time continued to hold up to scrutiny. In January the following year, after about 30 performances, hall president Junzo Fujimoto remarked how artists and audiences alike praised the hall highly. Cellist Yo-Yo Ma and conductor Kent Nagano went far as to label it the "world's best." In fact, I have come to think of the Tokyo Opera City project—after all that time, effort, and controversy—as the most successful collaboration of my career. The architect made very effort to understand the acoustical principles involved and to meet optimal requirements. My collaboration with Takenaka's acoustics group, too, could not have been more satisfactory.

Recognition came from an unexpected source—the Science Section of the *New York Times*. On Tuesday, April 19, 2000, a long article appeared under the "Art + Physics = Beautiful Music" (cribbed without permission, I confess, for the title of this chapter). The article includes photos of me and Yanagisawa in color. "An unusually intense collaboration between architects and acousticians," the writer noted, "has put the science of acoustics to test, with two major successes in Tokyo. The halls in question are the 1,632-seat concert hall of the multipurpose complex called Tokyo Opera City and the 1,810-seat opera house of the adjacent New National Theater." Yo-Yo Ma was quoted: "This [TOC] hall simply has some of the best acoustics in which I have ever had the privilege to play . . . a miracle." A number of drawings, in color, illustrated the influence of architectural features on the hall's acoustics.

Other Halls

Yet another acoustical project I worked on was the Tokyo Hamarikyu-Asahi Hall, in the fall of 1990. The chief architect was Makoto Usui. The drawings showed a small hall of 530 seats. I recommended that the stage size be reduced—its design was proportionately far too large for a hall that size—and that irregularities be added to the walls. After the tuning concert in August 1992, we added acoustical material in spaces above the stage, laid a thin carpet in the aisles, and placed architectural panels on the side walls above the balcony to reduce the overpowering high frequencies. This did the trick. Usui wrote me a year later to say that the hall had developed an excellent reputation.

At the same time as the NNT Opera House, I served as the acoustical consultant for the NNT Drama Theater, a hall with both proscenium and thrust stages. For the proscenium stage, the hall seats 1,010 and the pit accommodates about 72 musicians. At my first contact with the project, I expressed distress over the positioning of an enormous system of loudspeakers and stage lighting fixtures in the part of the ceiling nearest the proscenium. I advised that a reflective canopy was essential, like that in the opera house, for high speech intelligibility. Shocked to observe that the design showed the rear wall as a circular arc with the center of curvature at stage front, I explained that this was a sure recipe for a pronounced echo on stage and in the pit, as well as in the front rows of the audience.

I recommended quadratic residue diffusers on the rear wall to eliminate echoes, requiring a depth of about a foot. My advice was accepted and the QRDs installed. I later learned having won the battle to eliminate one of the light bridges in the ceiling of the TOC Opera House, Yanagisawa lost the battle for a reflecting surface above the Drama Theater proscenium. There was no way that the lighting and sound system people would let him prevail in both rooms.

Yanagisawa urged me to allow my name to be listed as acoustical consultant. I agreed, reluctantly, on the proviso that the opening night program booklet include the following notice: "This theater is unusual because emphasis has been placed on the thrust stage with a scenery loft situated above. Owing to this scenery loft, the acoustical enhancement of the performers' voices that would normally be obtained from a shaped, solid, ceiling must be supplied by an electronic sound system." I have no idea whether the notice was ever printed, but I doubt it.

The Drama Theater opened as an acoustical disaster. Critics complained that voices from the proscenium stage projected weakly and could not be understood. The theater management asked Hidaka to undertake a thorough acoustical examination with my assistance, to which I agreed at no added fee. When Hidaka approached me, he was quick to say that nobody thought I was in any way at fault for the acoustical failures. Hidaka, with my concurrence, went on to design a canopy—used for drama productions primarily—that could be hung as needed beneath the opening above the thrust stage and this seemed to be satisfactory.

Halls Outside the City

Starting in 1994, I worked on the acoustics for Mitaka City Hall. Yanagisawa was the architect once again, and Takenaka the contractor. A rectangular hall, it seats about 600. I did little to change the design presented at the outset. But at the tuning concert in August 1995, I detected a peculiar acoustical phenomenon—after a stop chord, the higher-frequency sounds seemed to rise quickly toward the stage ceiling. The defect was noticeable only when a single instrument or a small group such a trio was performing on stage. But when a full orchestra or a large group of musicians performed, the sound was acceptable. I proposed a minor modification to the end walls of the stage, which in no way would have disrupted its visual aesthetics, but the architect declined for budgetary reasons, and the hall remains today as it was in 1995.

My last acoustics job in Japan was a performance hall, Dai-ichi Seimei, sponsored by the Dai-ichi Mutual Life Insurance Company. In their own words, the hall was to be "the first private-sector facility of its kind whose operation is intimately tied to the community through the involvement of a specially established cultural-service nonprofit organization." It was intended expressly for concerts, with special emphasis on chamber music. Egg-shaped, with a stage at one end and a balcony that starts at the other end and extends over the sides of the stage, the hall was completed in November 2000, when Toshiyuki Okano took acoustical measurements. Much to our mutual surprise, he reported that a focusing effect along the entire centerline (front to rear) gave rise to what he called a "tunnel sound." Okano, alone, came up with the solution. Integral to the design, the rear wall had been covered with quadratic residue diffusers to eliminate the possibility of a rear-wall echo and, since the QRDs do not diffuse sound well at low frequencies, Okano's attention turned to them. He experimented with placing sound-absorbing materials selectively inside the channels of the diffusers, and found that this indeed solved the problem. All reports since opening in June 2001 have been positive.

This turned out to be my last consulting project anywhere. I felt I should retire because, going on 87, I was old enough to be my colleagues' grandfather—and in some cases, their great-grandfather. My experience in Japan had been favorable overall, with many successful outcomes, and it was especially satisfying to call it quits on the high note reached with the Dai-ichi Seimei performance hall. Of the Japanese halls I worked on, I consider the New National Theater Opera House and the Tokyo Opera City Concert Hall to be the best—near perfect—followed by the Dai-ichi Seimei hall, the Hamarikyu-Asahi Hall, and Mitaka City Hall in that order. I went on to write *Concert Halls and Opera Houses: Music, Acoustics, and Architecture* (Springer-New York Verlag, 2004), a collection, analysis, and synthesis of my varied experiences working on concert halls and opera houses over the years.

Epilogue

As I look back, my life seems like the images in a kaleidoscope—brightly colored, varied, and in constant motion. I lived through the greater part of the last century and faced transitions both large and small, fast and slow, better and worse. I was able to earn the entire wherewithal for my own college and graduate school education, even ending with a small cash surplus. World War II brought unique opportunities for a 26-year-old, which I made good use of through common sense, diligence, and boldness—some might say brazenness—to get things done in jig order. In every enterprise, my guiding principle has been to surround myself with others smarter than I—good people, I'm convinced, want to be in the company of equally good, or better, people. Through all this, I broadened my interests beyond mathematics, physics, and entrepreneurship, playing in a symphony orchestra and drumming in a dance band in my younger years and getting involved later on with arts and media management. But, as I quickly discovered, any organization is no better than its leader, who holds the key to success or failure. Whenever I did not take a forceful enough stand, as with the design of Philharmonic Hall in New York, the results were disastrous—or nearly so.

One central lesson I've learned is the value of taking risks and of moving on when those risks don't pan out or the odds look better elsewhere. Many may feel I went too far when I gambled my life's savings in a contest for ownership of a television station. The odds against success there were great, but the payoff appeared tens of times greater, so I threw caution to the winds. Using my off hours efficiently, I found the time to write 12 books and over 150 technical papers, always making sure I had something worth saying. And, as you have read here, I also found the time to pursue a number of other interests and activities orbiting my central endeavors—projects large and small, personal and professional, related and tangential.

An older professor at Harvard once told me that discovering something new was like falling down in the mud and coming up with diamonds. This wonderful image has stayed with me through the years—capturing in a nutshell the thrill of working toward a valuable end, with the prospect of almost limitless "ends" to follow. That thrill is still there for the taking—our supply of diamonds, just waiting to be pulled from the mud, is far from exhausted. There is so much still to learn about the earth and space, energy sources, how to change and augment genes, how to directly implant information in the brain, how to supply food, water, shelter, and health care for an exploding world population, and how to educate a population to compete in a world where the average level of knowledge and awareness is on the rise.

We must remain alert, too, for the unknown and the yet-to-be-known. As incredible as some new developments may seem, they are no more incredible than the prospect of sending and receiving images, sounds, and voices—anywhere on earth—with a wireless device smaller than a pack of playing cards would have seemed to nineteenth-century observers. The lesson here is that science, technology, and world events will continue to sweep us along a turbulent course of thrills, risks, and opportunities, a track that's almost always impossible to predict and that may require wholesale shifts in how we view the world and ourselves.

I often think of the people I've worked with over the years. Hardly a week goes by when I don't hear from a former colleague, student, or employee. Just the other day, I got a letter from George Kamperman, a staffer from my days at Bolt Beranek & Newman. "You are my mentor and one of the best friends in my lifetime," he wrote me. "You have had a profound positive impact on my life for half a century. . . . No matter what I was working on, you always made time to help me. You will always be in my heart." Each of us wishes to leave a legacy of some kind, and this is the best. What I cherish the most, and what I would most like to be remembered for, are the human connections I made in everything I've done.

Acknowledgments

I wish to thank, first of all, Philip N. Alexander of the Program in Writing and Humanistic Studies, Massachusetts Institute of Technology, for his extensive editorial assistance. My selection of material was influenced by courses I took on memoir writing at the Cambridge Center for Adult Education. My thanks also for the editorial suggestions that have flowed from a group of memoir writers meeting weekly at the home of Clark Abt in Cambridge. Unusual attention was given my manuscript by Marguerite Avery, Associate Editor, and Katherine Almeida, Senior Editor, at MIT Press. Copy editor Jeff Lockridge added continuity, punch, and clarity to an already well-worked text.

I am deeply grateful to my wife, Gabriella, for her constant encouragement, advice, and wonderful companionship.

Degrees, Awards, and Honors

1937, June 24. Harvard University, Master of Science in Engineering.

1939, February 23. Society of Sigma Xi. Elected as a Member of the Harvard Chapter.

1940, June 20. Harvard University (Universitas Harvardiana Cantabrigiae), Doctor of Science.

1944, May 12. Acoustical Society of America. Biennial Award of Merit (now called the R. Bruce Lindsay Award) for Contribution to Theoretical or Applied Acoustics by a Member of the Society Under 35 Years of Age.

1945, March 1. The United States of America, Office of Scientific Research and Development: Leo L. Beranek Has Contributed to the Successful Prosecution of the Second World War Through His Service in the National Defense Research Committee. (Electro-Acoustic Laboratory).

1945, March 1. The United States of America, Office of Scientific Research and Development: Leo L. Beranek Has Contributed to the Successful Prosecution of the Second World War Through His Service in the Office of Field Research. (Systems Research Laboratory).

1946, June. Cornell College (Iowa), Honorary Doctor of Science Degree.

1946–47. John Simon Guggenheim Fellow jointly at Massachusetts Institute of Technology and Harvard University.

1948, March 1. USA Presidential Certificate of Merit for outstanding fidelity and meritorious conduct in aid of the war effort in World War II (President Truman).

1951, April 21. Eta Kappa Nu Association. Appointed a Member for Attainments in the Profession of Electrical Engineering.

1952, January 1. Institute of Radio Engineers. Elected a Fellow.

1952, May 14. American Academy of Arts and Sciences. Elected a Fellow.

1957, December 6. Commonwealth of Massachusetts: Declared a Registered Professional Engineer.

1958, October 28. Acoustical Society of America: Elected a Fellow.

1961, November 10. Acoustical Society of America: Wallace Clement Sabine Award for Internationally Recognized Achievements in Acoustics. "For Twenty Years He Has Served His Field as Scientist, Teacher, Supervisor of Student Research, Author of Outstanding Books and Consultant to Architects."

1962. Made a Member of Phi Beta Kappa By Action of the Delta of Iowa at Cornell College.

1963, July 18. Province of Prince Edward Island: Declared a Registered Professional Engineer.

1965, May 14. Harvard University, Graduate School of Business Administration—Completion of Advanced Management Program, Spring 1965.

1966, April 15. National Academy of Engineering: Elected a Member.

1966, December 8. Le Groupement Des Acousticiens De Langue Francaise: Silver Medal.

1971, June. Worcester Polytechnic Institute: Honorary Doctor of Engineering Degree.

1975, April 9. Acoustical Society of America: Recipient of the 1975 Gold Medal Award.

1976. Radio and Television Commission of Southern Baptist Convention: Abe Lincoln Television Award (top USA award in television management).

1979, June. Suffolk University: Honorary Degree of Doctor of Commercial Science.

1982, June. Emerson College, Boston: Honorary Doctor of Laws Degree.

1983, May 17. Greater Boston Chamber of Commerce: Elected into the Academy of Distinguished Bostonians for Creative and Innovative Accomplishments, Community Service, and Sustained Leadership.

1984, June. Northeastern University: Honorary Doctor of Public Service Degree.

1987, January 9. Commonwealth of Massachusetts: Governor Michael S. Dukakis Declares This Day as Leo Beranek Day: "Leo Beranek Is a Tower of Strength in the Art Community."

1989, December 7. Cambridge Society for Early Music: The Arion Award for Contributions to Musical Organizations.

1994, May. American Academy of Arts and Sciences: President's Medal.

1994, June 7. Acoustical Society of America: Proclaimed Honorary Member for Contributions to Acoustics.

1996, May 14. National Council of Acoustical Consultants: Award of Appreciation for Scientific Advancements.

1996, October. Institute of Noise Control Engineering: Elected Distinguished Noise Control Engineer—Silver Bowl.

1997, June. Cornell College (Iowa): Honorary Life Trustee.

2000. American Academy of Arts and Sciences: Scholar-Patriot, Distinguished Service Award.

2000, May. American Institute of Architects: Honorary Member.

2002. International Institute of Acoustics and Vibration, Elected Honorary Fellow.

2002, December 4. Mexican Institute of Acoustics: Lord Rayleigh Medal for Advancement of Acoustics in Mexico.

2003, November 6. 2002 National Medal of Science, conferred by President George W. Bush, "For his leadership, dedication and contributions to the art and science of acoustics, for co-founding one of the world's foremost acoustical research and consulting firms, and for sustained contributions to scientific societies and civic organizations."

2004, November 15. American Society of Mechanical Engineers: The 2004 Per Bruel Gold Medal for Noise Control and Acoustics.

2004, October 6. Institute of Acoustics (England): Elected Honorary Fellow.

2007, September. Spanish Acoustical Society: top award, "Caracola de la Sea," for exceptional merits as a scientist and professor with a long and fruitful career in the field of acoustics.